AF894023

Stefanos Gkatzogiannis

Finite Element Simulation of Residual Stresses from Welding and High Frequency Hammer Peening

BAND 13

Versuchsanstalt für Stahl, Holz und Steine
Berichte zum Stahl- und Leichtbau

Finite Element Simulation of Residual Stresses from Welding and High Frequency Hammer Peening

by

Stefanos Gkatzogiannis

KIT Scientific Publishing

Karlsruher Institut für Technologie
Versuchsanstalt für Stahl, Holz und Steine

Finite Element Simulation of Residual Stresses from Welding and High Frequency Hammer Peening

Zur Erlangung des akademischen Grades eines Doktor-Ingenieurs von der KIT-Fakultät für Bauingenieur-, Geo- und Umweltwissenschaften des Karlsruher Instituts für Technologie (KIT) genehmigte Dissertation

von Dipl. Ing. Stefanos Gkatzogiannis

Tag der mündlichen Prüfung: 10. Juni 2020
Referent: Prof. Dr.-Ing. Thomas Ummenhofer
Korreferent: Prof. Dr. Andreas Taras

Impressum

KIT Scientific Publishing

Karlsruher Institut für Technologie (KIT)
KIT Scientific Publishing
Straße am Forum 2
D-76131 Karlsruhe

KIT Scientific Publishing is a registered trademark of Karlsruhe Institute of Technology.
Reprint using the book cover is not allowed.

www.ksp.kit.edu

This document – excluding parts marked otherwise, the cover, pictures and graphs – is licensed under a Creative Commons Attribution-Share Alike 4.0 International License (CC BY-SA 4.0): https://creativecommons.org/licenses/by-sa/4.0/deed.en

The cover page is licensed under a Creative Commons Attribution-No Derivatives 4.0 International License (CC BY-ND 4.0): https://creativecommons.org/licenses/by-nd/4.0/deed.en

Print on Demand 2022 – Gedruckt auf FSC-zertifiziertem Papier

ISSN 2198-7912
ISBN 978-3-7315-1066-6
DOI 10.5445/KSP/1000125591

Summary

The present doctoral dissertation focuses on the simulation of the residual stress state (RS) in steel weldments after their post-weld treatment with High Frequency Mechanical Impact (HFMI). Main goal is the establishment of an efficient engineering approach, which will include straightforward simulation models without compromising the accuracy, in order to estimate the RS field and enable a future evaluation of its influence on the fatigue life of the simulated components. The established approach should be applicable for both research and practical purposes. The Finite Element method is applied overall in the framework of the present study for the fulfilment of the research goal.

Prior to the application of HFMI, significant welding residual stresses (WRS), which cannot be neglected, are present in the as-welded state. Although research on welding simulation with the FE method is ongoing since decades, the methodology for practical applications remains vague. Based on existing knowledge, the present study establishes a straightforward engineering approach that considers all significant aspects for the accurate estimation of the WRS, which is proven applicable for several materials. Subsequently, the influence of several practical and special aspects of welding simulation on the simulated RS such as the applied values for thermal expansion coefficient, welding sequence, modelling of boundary conditions, phase changes, size of the modelled heat source etc. is investigated with this model. Validation of the simulations is based on direct comparison of the simulated temperature and RS profiles with respective experimental measurements found either in literature or in a completed research project. Conclusions, which can act as a modelling guide for the engineering practice, are presented. The commercial general-purpose FE software *ANSYS* has been applied for all welding simulations.

A similar approach is followed in the second part of the present research study, where modelling of HFMI is investigated. Once again, practical and special aspects of the HFMI simulation are considered. Amongst others, modelling of boundary conditions, density of HFMI treatment, scaling of components, applied values for the friction coefficient, material modelling and the influence of WRS are considered. Calibration of applied material models is carried out based on material testing from a completed research project and data from literature. Additionally, a series of drop tests for estimating the dynamic yielding behaviour of the investigated materials under the deformation mode present during HFMI treatment, are implemented for the first time. The present series of numerical investigations regarding HFMI either confirm or reject initial assumptions and conclusions from previous experimental and numerical studies. Additionally, new conclusions regarding the necessary aspects, which have to be considered or neglected, in order to achieve the desired accuracy, are proposed. The commercial general-purpose FE software *LS-DYNA* has been applied for all HFMI simulations.

Finally, recommendations for future work are presented regarding both the above-mentioned main parts of the present study and the numerical investigations regarding fatigue in general.

Kurzfassung

Die vorliegende Dissertation befasst sich mit der numerischen Simulation der Eigenspannungen von Schweißverbindungen nach ihrer Behandlung mit dem höherfrequenten Hämmerverfahren (HFH). Hauptziel ist die Entwicklung eines effizienten Inngenieurkonzeptes, das praxisorientierte Modelle einschließt, ohne die Genauigkeit der Ergebnisse zu beeinträchtigen. Es soll die Bestimmung des Einflusses der Eigenspannungen auf die Ermüdungsfestigkeit der simulierten Bauteile ermöglichen. Der entwickelte Ansatz sollte sowohl für Forschungszwecke als auch für praktische Zwecke anwendbar sein. In Rahmen der vorliegenden Studie wird die Finite-Elemente-Methode für die Erfüllung des aktuellen Forschungszieles angewendet.

Die Schweißverbindungen im wie-geschweißten Zustand haben schon vor der Nachbehandlung signifikante Schweißeigenspannungen. Obwohl die Forschung zum Thema Schweißsimulation schon seit Jahrzehnten betrieben wird, bleibt die Methodik für ihre praktische Anwendung unklar. Basierend auf dem vorhandenen Wissen wird in der vorliegenden Studie ein technischer Ansatz für die genaue Simulation des Lichtbogenschweißens entwickelt, der alle wichtigen Aspekte für die genaue Schätzung der Schweißeigenspannungen berücksichtigt und erwiesenermaßen für mehrere Materialien anwendbar ist. Mithilfe dieses Modells wurde der Einfluss einiger praktischer oder spezieller Aspekte der Schweißsimulation, wie z.B. die verwendeten Werte des Wärmeausdehnungskoeffizienten, die Schweißreihenfolge, die Modellierung der Randbedingungen, die Phasenumwandlungen, die Abmessungen der Wärmequelle usw. auf die gerechneten Eigenspannungen untersucht. Die Validierung der Simulationen basiert auf einem direkten Vergleich der simulierten Temperaturprofile und Eigenspannungen mit entsprechenden experimentellen Messungen, die entweder in der Literatur oder in einem abgeschlossenen Forschungsprojekt gefunden wurden. Es werden Schlussfolgerungen präsentiert, die als Modellierungsleitfaden für die Ingenieurpraxis dienen können. Die kommerzielle FE-Software *ANSYS* wurde für alle Schweißsimulationen angewendet.

Ein ähnlicher Ansatz wird im zweiten Teil der vorliegenden Doktorarbeit verfolgt, in der die FE-Modellierung des höherfrequenten Hämmerns HFH untersucht wird.

Auch hier werden praktische und spezielle Aspekte der HFH-Simulation betrachtet. Unter anderen werden berücksichtigt: die Modellierung der Randbedingungen, die Überlappung von HFH-Schlägen, die Skalierung von Bauteilen in der Simulation, die angewendeten Werte für den Reibungskoeffizienten, die Werkstoffgesetze und der Einfluss der Schweißeigenspannungen. Die Kalibrierung der verwendeten Werkstoffgesetze erfolgt mithilfe von der Werkstoffcharakterisierung aus einem abgeschlossenen Projekt und von Daten aus der Literatur. Darüber hinaus wurde eine Reihe von Fallversuchen für die Bestimmung der dynamischen Streckgrenze von den untersuchten Werkstoffen durchgeführt. Die aktuelle Serie von numerischen Untersuchungen des HFH bestätigt entweder oder widerlegt Anfangsannahmen und Schlussfolgerungen aus früheren experimentellen und numerischen Untersuchungen. Zusätzlich werden neue Schlussfolgerungen bezüglich der notwendigen Aspekte vorgeschlagen, die zu berücksichtigen sind, um die gewünschte Genauigkeit zu erreichen. Die kommerzielle FE-Software *LS-DYNA* wird für alle HFH-Simulationen angewendet.

Abschließend werden Empfehlungen für zukünftige Untersuchungen zu den oben genannten Hauptthemen der vorliegenden Doktorarbeit sowie zum Gebiet Ermüdung und FE-Simulationen im Allgemeinen gegeben.

...«Κι ἂν πτωχικὴ τὴν βρῇς, ἡ Ἰθάκη δὲν σὲ γέλασε.
Ἔτσι σοφὸς ποὺ ἔγινες, μὲ τόση πείρα,
ἤδη θὰ τὸ κατάλαβες ἡ Ἰθάκες τί σημαίνουν.»

..."And if you find her poor, Ithaka won't have fooled you.
Wise as you will have become, so full of experience,
you'll have understood by then what these Ithakas mean."

Ἰθάκη (Ithaka)

by C. P. Cavafy,

translated by Edmund Keeley

metallurgy

From French métallurgie, from Ancient Greek μεταλλουργός (metallourgós, "worker in metal"), from μέταλλον (métallon, "metal") + ἔργον (érgon, "work").

To my parents Genovefa and Stylianos for unveiling to me the fathomless beauty of knowledge, for their love and support.

Preface

During my undergraduate studies at the School of Civil Engineering, at the Aristotle University of Thessaloniki, I was fascinated by two subjects of structural engineering, which would shape my academic future: steel structures and computational mechanics. The possibilities of the accurate design enabled by the mechanical properties of steel, its application on more complicated structures and the fact that during design of such structures, the challenge of combining complex scientific knowledge with practical engineering solutions arises, have been the reasons for concentrating on the first subject. Regarding the latter one, the computational mechanics, it has been from the beginning pure fascination in front of the capabilities, which are enabled by introduction of computers in modern engineering. Therefore, when I decided to pursue the doctoral title of engineering it was the natural course of events to land on the field of simulation of steel structures.

When I first met Professor Thomas Ummenhofer in June 2012 and made my intentions of having a doctoral dissertation regarding numerical analysis in the field of steel structures known to him, he proposed the present subject. Back then, my knowledge on welding was restricted to design of weldments against static loads and the terms "post-weld treatment" and "residual stresses" were unknown to me. I agreed to begin my postgraduate research in *Karlsruhe Institute of Technology* on this subject without imagining the very interesting journey, which was about to begin. Over the years, I investigated several aspects, which were not strictly attached to the main subject, they were not dots of a straight path, but small sidesteps I had to take. I strongly believe now that they helped me significantly to better comprehend the subjects of plasticity, residual stress, metallurgy and fatigue amongst others. The completion of the present study broadened my scientific knowledge but more significantly, it reshaped my personality by increasing my work ethics on a level I could not foresee.

I would sincerely like to thank my supervisor Professor Thomas Ummenhofer for giving me the opportunity to work in such a challenging environment, like the *KIT Steel and Lightweight Structures Institute - Research Centre for Steel, Wood and Masonry*. Without his trust in me and my previous education, this dissertation would not have been realized. His scientific vigour was and will always be an inspiration for me. His ability to lead the younger researchers with such insightfulness on so many different subjects of our field will always be for me a source of admiration, something to look up to.

Furthermore, I would like to thank my colleagues, scientific and laboratory associates of the *KIT Steel and Lightweight Structures Institute,* for the smooth cooperation over the years. Especially I have to mention Philipp Ladendorf for the interesting scientific discussions and his input regarding material testing over the years, my research assistant Ioannis Savvanidis for his contribution in designing and setting up the drop tests of the present study and Dr. Tim Zinke for his valuable advice regarding the process of writing and publishing my doctoral dissertation. Special thanks to Dr. Majid Farajian and Mr. Jan Schubnell for the fruitful collaboration in the framework of the project *HFH-Simulation*. Without their input, the present study could not have been finalized. For their financial support to the same research project, which provided significant input for the present dissertation, I should acknowledge as well the *DVS – Deutscher Verband für Schweißen und verwandte Verfahren* and the *Arbeitsgemeinschaft industrieller Forschungsvereinigungen "Otto von Guericke" e.V. (AiF).*

I would like to thank Professor Andreas Taras for being the second reviewer of the present dissertation, the time he invested for this task, his valuable remarks and overall for his positive feedback. I thank him as well along with Professor Peter Betsch and Professor Joachim Blaß for being members of the examining committee and for acknowledging the hard effort for completing this dissertation and the quality of my work.

At this point, I would like to state explicitly my gratitude to my mentor Professor Peter Knoedel, a great scientist and a wonderful person, for his patience, for all the things he taught me and for all the memorable moments. He never stopped motivating me take up new challenges. He has been rigorous but just, he kindly pinpointed my mistakes and cheerfully congratulated my successes, always respecting my personality. For me he is much more than a tutor, a director or a colleague, he has become a heartwarming, dearest friend. For all these, Peter, thank you.

I should not forget to mention my friends Alexandros, Andreas, Georgios, Ioannis, Michalis, Orestes and Vasilis, with whom I lived over the last years many memorable moments, which helped me going. Many special thanks to my companion Aikaterini, for her support and for the nice moments we had in Karlsruhe. Her sweet, vivid and kind personality has been a shelter in difficult moments. The healthy competition with her wit and scientific duality have pushed me to improve myself significantly. Additionally, I would like to thank her for contributing to the syntax of the *MATLAB* code for the present study. Finally yet importantly, I would like to thank my parents Genovefa and Stylianos to whom I devote this dissertation. Without their love and support, this doctoral dissertation would not have been completed. I am grateful to them for showing me the joy of knowledge, for teaching me to choose always the hard path of virtue and for making me understand in their very own words, what the Ithacas of this world mean.

Stefanos Gkatzogiannis

Karlsruhe, September 2020

Table of Contents

List of Publications

Publications Carried Out in the Framework of the Present Dissertation

The following studies were published in the framework of the present doctoral dissertation and acted as milestones during its realization. They are presented categorized and in chronological order:

Publications in peer reviewed academic and technical journals

Gkatzogiannis S., Knoedel P., Ummenhofer T.; Calibration of HFMI Simulation based on Drop Tests, *Journal of Material Engineering and Performance*, published online on 04 May 2020.

Gkatzogiannis S., Knoedel P., Ummenhofer T.; Strain Rate Dependency of Simulated Welding Residual Stresses, *Journal of Material Engineering and Performance* **27** (10), pp. 5079-5085, 2018.

Knoedel P., Gkatzogiannis, S., Ummenhofer T.; Practical Aspects of Welding Residual Stress Simulation, *Journal of Constructional Steel Research* **132**, pp. 83-96, 2017.

Publications in peer reviewed conference proceedings

Gkatzogiannis S., Knoedel P., Ummenhofer T.; FE Simulation of High Frequency Mechanical Impact (HFMI) Treatment – First Results, *Proceeding of NORDIC STEEL 2019, The 14th Nordic Steel Construction Conference, Copenhagen 18–20 September 2019, ce/papers* **3** (3-4), pp. 797-802, Ernst & Sohn, *2019.*

Gkatzogiannis S., Knoedel P., Ummenhofer T.; A Pseudothermal Approach for Simulating the Residual Stress Field Caused by Shot Blasting, *Proceedings of the VIII International Conference on Computational Methods for Coupled Problems in Science and Engineering, Sitges 3-6 June 2019,* pp. 777-788, 2019.

Gkatzogiannis S., Knoedel P., Ummenhofer T.; Simulation of Welding Residual Stresses – From Theory to Practice, Selected Peer Reviewed Papers from the *12th International Seminar Numerical Analysis of Weldability, Graz – Schloss Seggau 23-26 September 2018,* published in Sommitsch C., Enzinger N., Mayr P.*; Mathematical Modelling of Weld Phenomena* **12***,* pp. 383-400, 2019.

Gkatzogiannis S., Knoedel P., Ummenhofer T.; Reviewing the Influence of Welding Setup on FE-simulated Welding Residual Stresses, *Proceedings of the 10th European Conference on Residual Stresses - ECRS10, Leuven 11-14 September 2018,* published in *Materials Research Proceedings* **6**, pp. 197-202, 2018.

Gkatzogiannis S., Knoedel P., Ummenhofer T.; FE Welding Residual Stress Simulation – Influence of Boundary Conditions and Material Models, *Proceedings of EUROSTEEL 2017, Copenhagen 13–15 September 2017, ce/papers* **1**, Ernst & Sohn, 2017.

Gkatzogiannis S., Knoedel P., Ummenhofer T.; Influence of Welding Parameters on the Welding Residual Stresses, *Proceedings of the VII International Conference on Coupled Problems in Science and Engineering, Rhodes 12–14 June 2017*, pp. 767–778, 2017.

Knoedel P., Gkatzogiannis S., Ummenhofer T.; FE Simulation of Residual Welding Stresses: Aluminum and Steel Structural Components, selected peer reviewed papers from the *13th International Aluminium Conference INALCO 2016, Naples 21–23 September 2016*, published in *Key Engineering Materials* **710**, pp. 268-274, 2016.

Conference presentations

Gkatzogiannis S., Knoedel P., Ummenhofer T.; FE Simulation of the HFMI Treatment - Previous and Upcoming Results, *Symposium Mechanische Oberflächenbehandlung 2019 – 8th Workshop Machine Hammer Peening, Karlsruhe 22-23 October 2019.*

Gkatzogiannis S., Knoedel P., Ummenhofer T.; Calibration of HFMI Simulation based on Drop Tests, *EUROMAT 19, Stockholm 1-5 September 2019.*

Schubnell J., Carl E., Farajian M., Gkatzogiannis S., Knödel P., Ummenhofer T., Wimpory R., Eslami H.; Residual Stress Relaxation in HFMI-Treated Fillet Welds After Single Overload Peaks, *IIW Commission XIII, Fatigue of Welded Components and Structures XIII-2829-19,* 2019.

Gkatzogiannis S., Knoedel P., Ummenhofer T.; Strain Rate Dependency of Weld Simulation, *EUROMAT 17, Thessaloniki 17-22 September 2017.*

Knoedel P., Gkatzogiannis S., Ummenhofer T.; Creep-behaviour of Welded Structures, *Simulationsforum 2016 – Schweißen und Wärmebehandlung, Weimar 8-10 November 2016*, pp. 209–219, 2016.

Research projects:
Schubnell J., Gkatzogiannis S., Farajian M., Knoedel P., Luke T., Ummenhofer T.; *Rechnergestütztes Bewertungstool zum Nachweis der Lebensdauerverlängerung von mit dem Hochfrequenz-Hämmerverfahren (HFMI) behandelten Schweißverbindungen aus hochfesten Stählen*, Abschlussbericht DVS 09069 – IGF 19227 N, Fraunhofer Institut für Werkstoffmechanik, Freiburg und KIT Stahl- und Leichtbau, Versuchsanstalt für Stahl, Holz und Steine, Karlsruhe, 2019.

Publications in Regard to Fatigue and HFMI

The following studies were published parallel to the present doctoral dissertation. They are mentioned at this point, categorized and in chronological order, as they are relevant to the general subject of HFMI and fatigue:

Publications in peer reviewed academic and technical journals

Schubnell J., Carl E., Farajian M., Gkatzogiannis S., Knödel P., Ummenhofer T., Wimpory R., Eslami H.; Residual Stress Relaxation in HFMI-Treated Fillet Welds after Single Overload Peaks, *Welding in the World* **64**, pp. 1107–1117, 2020.

Gkatzogiannis S., Weinert J., Engelhardt I., Knoedel P., Ummenhofer T.; Correlation of Laboratory and Real Marine Corrosion for the Investigation of Corrosion Fatigue Behaviour of Steel Components, *International Journal of Fatigue* **126**, pp. 90-102, 2019.

Weinert, J., Gkatzogiannis, S., Engelhardt, I., Knödel, P., Ummenhofer, T.; Erhöhung der Ermüdungsfestigkeit von geschweißten Konstruktionsdetails in korrosiver Umgebung durch Anwendung höherfrequenter Hämmerverfahren, *Schweißen und Schneiden* **70** (11), pp. 782–789, 2018.

Publications in peer reviewed conference proceedings:
Ummenhofer, T., Gkatzogiannis, S., Weidner, P.; Einfluss der Korrosion auf die Ermüdungsfestigkeit von Konstruktionen des Stahlwasserbaus, *Tagungsband BAW Kolloquium - Korrosionsschutz und Tragfähigkeit bestehender Stahlwasserbauverschlüsse, Karlsruhe 8-9 Februar 2017,* pp. 80-86, 2017.

Conference presentations:
Schubnell J., Carl E., Farajian M., Gkatzogiannis S., Knödel P., Ummenhofer T., Wimpory R., Eslami H.; Residual Stress Relaxation in HFMI-Treated Fillet Welds After Single Overload Peaks, *Symposium Mechanische Oberflächenbehandlung 2019 – 8th Workshop Machine Hammer Peening, Karlsruhe 22-23 October 2019.*

Weinert J, Gkatzogiannis S., Engelhardt I., Knoedel P., Ummenhofer T.; Application of High Frequency Mechanical Impact Treatment to Improve the Fatigue Strength of Welded Joints in Corrosive Environment, *IIW Commission XIII, Fatigue of Welded Components and Structures XIII-2781-19*, 2019.

Weinert J, Gkatzogiannis S., Engelhardt I., Knoedel P., Ummenhofer T.; Potential der Schweißnahtnachbehandlung mithilfe von höherfrequenten Hämmernverfahren für den Einsatz an Offshore Gründungsstrukturen, *19. Tagung Schweißen in der Maritimen Technik und im Ingenieurbau, Hamburg 24-25 April 2019,* pp. 92-105, 2019.

Gkatzogiannis S., Weinert J., Engelhardt I., Knoedel P., Ummenhofer T.; Corrosion Fatigue Behaviour of HFMI-Treated Welded Joints of Steel S355 – Correlation of Testing Methods, *EUROMAT 17, Thessaloniki 17-22 September 2017.*

Weinert J., Löschner D., Gkatzogiannis S., Engelhardt I., Knödel P., Ummenhofer T.; Influence of Seawater Corrosion on The Fatigue Strength of High Frequency Hammerpeened (HFH-Treated) Welded Joints, *Joint European Corrosion Congress 2017, EUROCORR 2017 and 20th International Corrosion Congress and Process Safety Congress 2017; Prague 3-7 September 2017.*

Research projects:
Ummenhofer T., Engelhardt I., Knoedel P., Gkatzogiannis S., Weinert J., Loeschner D.; *Erhöhung der Ermüdungsfestigkeit von Offshore-Windenergieanlagen durch Schweißnahtnachbehandlung unter Berücksichtigung des Korrosionseinflusses*, Schlussbericht, DVS 09069 – IGF 18457 N, KIT Stahl- und Leichtbau, Versuchsanstalt für Stahl, Holz und Steine, Karlsruhe und Hochschule für angewandte Wissenschaften München, Labor für Stahl- und Leichtmetallbau, 2018.

List of Figures

List of Abbreviations

Al-T:	Aluminium Total
AY:	Austenite
BA:	Bainite
BM:	Base Material
B-T:	Boron Total
CG:	Coarse Grained
CMM:	Cooling Material Model
FE:	Ferrite
FE:	Finite Element
FZ:	Fusion Zone
HAZ:	Heat-Affected Zone
HFH:	Hochfrequentes Hämmernverfahren
HiFIT:	High Frequency Impact Treatment
IWM:	Fraunhofer Institute for Mechanics of Materials (IWM), Freiburg, Germany
MA:	Martensite
PE:	Pearlite
PIT:	Pneumatic Impact Treatment
PM:	Parent Material

RS: Residual Stresses

STAAZ: Spitzentemperatur, Austenitisierungs- und Abkühlungszeit

TRIP: Transformation Induced Plasticity

UFG: Ultra Fine Grained

WPS Welding Procedure Specifications

WRS: Welding Residual Stresses

List of Notations

Symbol	Unit	Meaning
δ	m	indentation depth
F	N	force
ν	-	Poisson's ratio
E	Pa	Young's Modulus
R	m	radius
p_m	Pa	average contact pressure
σ_y	Pa	static yield stress
v	m/s	impact velocity
e	-	coefficient of restitution
σ_y'	Pa	dynamic yield stress
V	m^3	volume
q	W/m^3	power density
a, b, c	m	width, depth and length of the ellipsoidal quadrants of Goldak's heat source
f_r	J	heat fraction deposited in the rear ellipsoidal quadrant of Goldak's heat source
f_f	J	heat fraction deposited in the front ellipsoidal quadrant of Goldak's heat source
Q	J/s	effective heat input rate

C	m	characteristic radius of flux distribution
v	m/s	welding source travel
t	s	time
τ_{lag}	-	lag factor ("phase shift") needed to define the position of weld heat source at time t = 0
V	V	voltage
I	A	current of the weld metal arc
η	-	weld metal arc efficiency
ρ	kg/m^3	density
c_{heat}	J/(kg · K)	specific heat capacity
T	K or °C [1]	temperature
K_{xx}, K_{yy}, K_{zz}	W/(m · K)	thermal conductivity in the x, y, and z directions
$\dddot{q}$	W/m^3	heat generation rate per unit volume
v_x, v_y, v_z	m/s	velocity for mass transport of heat in x, y, and z directions, respectively
ΔU	J	internal energy of a closed system
Q_E	J	heat supplied to a closed system
W	J	amount of work produced into a closed system
q	W/m^3	local heat flux density
K	W/(m · K)	local materials conductivity

[1] equations of physical laws are expressed in K unless otherwise mentioned, while data for material properties are given in °C in order to offer a better overview to the reader

q/A	J/s	heat flow through a surface
h_C	$W/(m^2 \cdot K)$	convective coefficient of heat losses
h_R	$W/(m^2 \cdot K)$	radiative coefficient of heat losses
T_s	K	temperature of the surface of heat convection
T_b	K	temperature of the surrounding fluid
h_T	$W/(m^2 \cdot K)$	total coefficient of heat losses
$A1$	°C	static start temperature for austenitization
$A3$	°C	static completion temperature for austenitization
$A3'$	°C	dynamic completion temperature for austenitization
n_i	-	proportion of a microstructure phase
n_i^{eq}	-	equivalent proportion of a microstructure phase
ϑ	-	coefficient of Leblond's model for austenitization
n_i^{mart}	-	proportion of a microstructure transformed to martensite
M_s	°C	start temperature for martensitic transformation
T_q	°C	quenching temperature
T_{max}	°C	maximum temperature in a thermal cycle
t_{85}	s	cooling time from 800 °C to 500 °C
t_a	s	time above asutenitazation temperature *A1*
d_{aust}	m	austenite grain size
$\mathbf{fs(u)}$	N	internal force vector, function of deformation

$\mathbf{p}(t)$	N	external load vector for a system of finite elements as a function of time
ε^{th}	-	thermal strains
α^{se}	-	secant coefficient of thermal expansion
H	Pa	tangent modulus
$f(\sigma)$	-	yield function
J_1, J_2, J_3	N^2/mm^4	1st, 2nd and 3d invariant of stresses
τ	N/mm^2	yield stress in shear
$\dot{\hat{\varepsilon}}_{pl}$	s^{-1}	equivalent plastic strain rate
α, β, γ and δ	-	coefficients with no direct physical meaning characterizing the strain rate hardening behaviour
$\mathbf{m}$	kg	mass matrix
$\mathbf{c}$	$(N \cdot s)/m$	damping matrix
$\mathbf{u}, \dot{\mathbf{u}}, \ddot{\mathbf{u}}$	m	displacement, velocity and acceleration vector
$\sigma_F{}'$	Pa	von Mises flow stress
$\varepsilon_p{}'$	-	equivalent plastic strain
A	Pa	coefficient of the Johnson-Cook model
B	Pa	coefficient of the Johnson-Cook model
Γ	-	coefficient of the Johnson-Cook model
ζ	-	coefficient of the Johnson-Cook model
ϑ	-	coefficient of the Johnson-Cook model
$\dot{\varepsilon}$	-	reference strain rate

$\dot{\varepsilon}'$	-	investigated strain rate
T_o	°C	reference temperature level
T_m	°C	investigated temperature level
m_{SRS}	-	strain rate sensitivity
α_R	1/kg	coefficient of the Rayleigh model
β_R	1/Pa	coefficient of the Rayleigh model
f	Hz	frequency
ξ	-	damping ratio
Ds	-	damping factor
F_f	N	friction force
μ	-	friction coefficient
F_N	N	normal force exerted by each surface on the other, perpendicular to the contact sufrace
$v_{sliding}$	m/s	relative sliding velocity between two surfaces
D	m	diameter

1 Introduction

1.1 Problem Statement

It was more than 50 years past the first patented application of welding in Russia at the end of the 19th century [27], when engineers started to realize the phenomenon of fatigue fracture in weldments. Events like the collapses of the *Point Pleasant Bridge* in the US and the *Alexander Kielland* offshore platform in Norway, which were caused due to fatigue cracking of welded connections and led to losses of human lives [66], increased the awareness regarding fatigue design and exhibited the vulnerability of welded joints against cyclic loading[1]. Ever since, methods and recommendations regarding fatigue design ([35], [76] e.g.), steel quality ([29] e.g.), welding quality ([79] e.g.) and non-destructive testing ([80] e.g.) have been developed and activated respectively. Therewith, the fatigue life of steel structures can be predicted with safety, the ductile performance of the parent material and the welded joint are assured and joining defects can be avoided or detected.

Nonetheless, welded joints remain the Achilles heel of steel structures, when they are subjected to fatigue loading. The fatigue strength of welds lies significantly lower than that of parent material due to the notch effect and the respective concentration factor, the tensile welding residual stresses (WRS)[2], the unavoidable welding defects and the reduced ductility of the heat-affected zone (HAZ). Hence, extending fatigue life of welded joints leads to significant increase of a construction's life cycle.

Several methods have been developed in the last decades with the purpose of increasing fatigue life of welds, with *High Frequency Mechanical Impact* treatment[3] (HFMI) [118] being one of the most straightforward and effective (see [167]). It can be applied through the use of a device by the craftsman or by a robot

---

[1] the problem of fatigue regarding parent (unwelded) metallic materials was already known from the 19th century, worth mentioning are the Versailles rail accident and the work of Julius Albert and August Wöhler

[2] a list of abbreviations is given at the begining of the present dissertation at page xxxiii

[3] or Hochfrequetes Hämmerverfahren (HFH) in German

both during manufacturing process and on existing and new structures in the field. Therewith, a significant increase of fatigue strength of even more than 100 % in some cases is possible (see [118]). The first HFMI application was designed in the 70's in the Soviet Union under the name Ultrasonic Impact Treatment (UIT) [153]. Nowadays, two different types of HFMI devices are manufactured in Germany, *High Frequency Impact Treatment* (abbreviated as *HiFIT*, described in [161] and [162]) and *Pneumatic Impact Treatment* (abbreviated as *PIT* or *PITec*, described in [98]). The principle remains the same in all cases. A pin made of hard steel is accelerated towards the weld toe by an appropriate mechanism, which differs though for each manufacturer. The weld toe is plastically deformed, the weld notch effect is reduced and compressive residual stresses (RS), which counterbalance the detrimental tensile WRS, are introduced in the treated area [167]. *HiFIT* and *PITec* devices are presented in Figure 1[4].

(a)

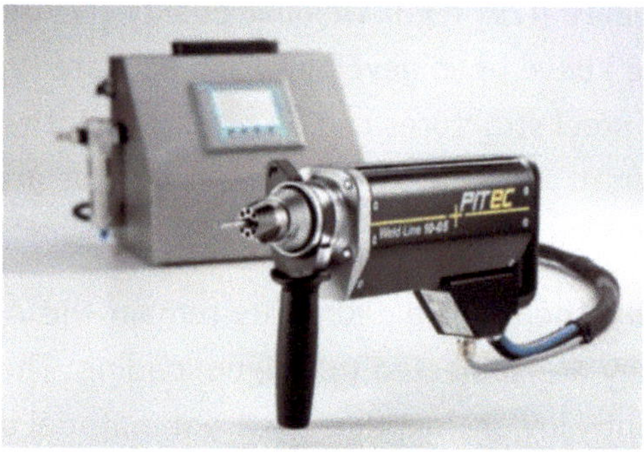

(b)

Figure 1: HFMI devices manufactured in Germany: a) *HiFIT* (courtesy of *HiFIT* GmbH); (b) *PITec* (courtesy of *PITec* GmbH)

The effectiveness of HFMI for the extension of fatigue life of welded joints has been thoroughly validated by experimental investigations in the past (see [167], [168], [171], [181], [182], [183] etc.). The fatigue life of HFMI-treated butt-welds has been proven to be higher than that of parent material in [109], [167], [168], [173] etc. The effectiveness of the method was also investigated for different geometries, plate thicknesses and parent material: transversal and longitudinal stiffeners and hollow sections of various thicknesses were tested in [33], [109],

[4] figures of the present manuscript are referenced in the text as Figure

[167], [171], [175], [176] and [185]. In most cases the inverse slope *m* of the calculated SN curve of the HFMI-treated specimens was larger than 5 [33], [167], [163], [171] and a clear increase of fatigue strength was documented. Tests on specimens made of S355, S690, S910, S1100 and S1300 ([33], [167], [168], [169], [171], [174], [176] etc.) have displayed a dependency of the HFMI effectiveness on the yield strength of the investigated material, with high strength steels displaying more potential. The higher introduced compressive RS are to be accounted for this increase in effectiveness. It is clear from the above, that the HFMI enhancement of fatigue strength depends on many parameters and respective factors for the consideration during the design of the increased fatigue life have been proposed in [119], [172].

Extensive research on HFMI during the last two decades enabled the regulation of the method by the *International Institute of Welding* (IIW) according to [118] by analogy to existing recommendations for as-welded specimens (see [35], [76]). Influence of material nominal yield strength and fatigue loading stress ratio was as well taken into consideration. Nevertheless, the approach of SN curves and the respective proposed FAT classes in both cases are quite conservative: the 95 % confidence interval is proposed as the characteristic fatigue strength of each investigated notch detail. Moreover, the proposed FAT classes are calculated based on several test series carried out by different research groups on welded specimens, which are nominally identical, but in reality can qualitatively differ significantly from each other. This problem is thoroughly described in [38]. Although this approach is reasonable enough, when fatigue design recommendations for the practitioner have to be compiled, extracted FAT classes can be too conservative for weldments of high quality.

Numerical modelling of HFMI could be a valuable alternative to costly fatigue tests. Coupled with weld simulation it could enable a safe prediction of the RS field, taking into consideration the various unique parameters of each investigated case, such as welding parameters, notch effect, complex geometries, material etc. The calculated WRS field should be input for an accurate calculation of fatigue life through simulation of RS from HFMI. Computational welding mechanics (see [60], [111]) have evolved rapidly in the last decades and results with satisfying precision regarding WRS and respective deformation can be extracted [57]. Some numerical investigations of HFMI have been carried out during the last years as well, neglecting however in most cases significant effects of the process (see, [1], [44], [68], [98], [99], [103], [142], [150], [167], [187], etc.). For example, precise modelling of material behaviour, as a high-speed impact event is simulated, movement of the HFMI pin, boundary conditions as well as WRS is required. These aspects were taken into consideration only in very recent studies, which were published parallel and shortly prior to the conclusion of the present dissertation [36], [108].

Objective of the present study is the establishment of a validated engineering approach for the simulation of HFMI, taking into consideration all the significant aspects of the process, in order to provide a robust prediction of the introduced RS. The developed approach should serve a dual role. It should be applicable both for research purpose, whereby it could be used as a tool for sensitivity analyses and in extension further investigation and improvement of the method, and in practice, in order achieve less conservative design. During the development of the presented method, all predominant factors that affect RS should be considered. The application of the method should be straightforward, meaning that special knowledge of physical metallurgy apart from the basic knowledge of material science taught to undergraduate level of engineering would not be required, without compromising though the preciseness of the results.

1.2 Research Methodology

The subject of the present study can be divided into two major fields, the weld process and the HFMI simulation. Although the main subject of the presented study is numerical, analytical calculations as part of the method were deployed, when it was considered necessary. Moreover, experimental results, which were extracted in the framework of the present or others studies ([3], [21], [146]) were used as input for the developed approach or for validation of the results.

Several investigations regarding the first field have been presented in a series of previous publications by the author, which were carried out in the framework of the present doctoral dissertation. A straightforward, appropriate for practical applications approach for modelling the WRS was developed [92] and validated based on measurements found elsewhere [1], [21]. The method was extended in

[20], [55] and [56] for simulating various materials and used for sensitivity analyses regarding parameters of material modelling [53], welding parameters [56] and boundary conditions [55], [57]. The model was adapted and revalidated for the presently investigated materials, based on measurements from the research project *HFH-Simulation* [146]. Whenever possible, values from literature were applied for common material properties. As the above-mentioned publications were carried out in the framework of the present doctoral dissertation, the main aspects of the method are thoroughly presented in this study as well, when necessary. Commercial FE software *ANSYS* was applied for all numerical investigations of weld processes [4].

Material data, which was used as input for the HFMI simulation, was extracted from drop tests. The results of the drop tests were evaluated based on appropriate analytical and numerical calculations and were compared with respective experimental results for the same batches of the investigated materials from *HFH-Simulation* [146]. In this case as well, values from literature were applied for common material properties, when it was considered that the preciseness would not be compromised. Results from a previous study, wherein the HFMI treatment of an unnotched plate was investigated, were used for a first step validation of the developed HFMI modelling approach [44]. After the HFMI simulation model was validated, it was coupled with the welding simulation model and the results were once again validated based on the measurements from the research project *HFH-Simulation* [146]. During the development and validation of the present approach, significant conclusions regarding practical aspects of the HFMI simulation were drawn. Material modelling, definition of boundary conditions, simulation of real scale fatigue tests and modelling techniques regarding the motion of HFMI pin were investigated amongst others. The commercial FE software *LS-DYNA* was applied for all numerical investigations of the HFMI treatment [114].

As it was mentioned above, the developed approach should be appropriate for both research and practical applications. Therefore, an appropriate balance between preciseness and computational effort should be held at all times. Ultimate objective of the present study was to enable a safe prediction of the RS in the areas of the components, which are susceptible to fatigue cracking i.e. the near-surface region of the heat-affected zone, in order to allow for a safe estimation of fatigue life. An empirical thumb rule of a deviation equal to ± 10 % of the investigated material's yield stress or smaller between simulated and measured RS at these areas of the components was considered sufficient and feasible for the targeted engineering application and was fulfilled in most of the investigated cases.

1.3 High Performance Computing

The high performance research computer ForHLR I of the *Steinbuch Centre for Computing* at the *Karlsruhe Institute of Technology* was applied for carrying out the numerical investigations of the present doctoral dissertation. In the case of the welding simulations in *ANSYS* [4], a fat node was used for Parallel Processing with 16 processors. Initial memory request was set to 500 GB in each case. To give an example, the duration of simulating a 1000 mm x 370 mm x 10 mm, 4-pass fillet weld, meshed with 175827 nodes and 160296 elements using these computing resources was in real time 97 hours 45 min and 34 seconds. In the case of the HFMI simulation, 16 nodes with 16 processors each were applied with Massive Parallel Processing (MPP). For simulating the 3-pass treatment of a component with dimensions of 20 mm x 20 mm x 10 mm, meshed with 195640 nodes and 208022 elements, 7 hours 21 minutes and 52 seconds elapsed. Deployment of up to 512 processors in total was the upper limit regarding HFMI simulations, due to the available number of *LS-DYNA* licenses [113].

1.4 Outline of the Present Dissertation

The present dissertation is organised in 7 chapters, the present introductory one and seven more. In the 2nd chapter, a thorough review of the theoretical background for the present analytical and numerical investigations is made. The analytical investigations were carried out in order to evaluate the above-described drop tests. The theoretical background of all numerical investigations, which are necessary for the calculation of RS from welding and HFMI is presented as well. The highlights of previous work by other authors and the author of the present dissertation are exhibited, so that a comprehensive overview of the state of the art regarding the present subject becomes available to the reader.

In the 3rd chapter the numerical investigations regarding residual stresses from welding, which were carried out in the framework of the present study, are presented. Mesh and modelling restrictions are discussed. After a detailed analysis of the applied methodology, results regarding three different weldments are presented. A series of investigations, regarding the influence of various practical and special aspects of weld simulation, like boundary conditions, material modelling and welding parameters etc. on the modelled WRS is presented. Conclusions regarding the welding simulation and recommendations regarding future work are summarized at the end of the chapter.

The 4th chapter reports on the experimental investigations, which were carried out in the framework of the present study, along with the analytical and numerical models, which were applied for the evaluation of the test results. The test set up is described thoroughly and restrictions and errors that arise are reported. The inevitable assumptions for the simplifications of the analytical model are highlighted. The test results and the extracted material properties are presented and compared with respective results from other sources.

In the 5th chapter, the numerical study on HFMI and the introduced RS is described. The methodology and the results of some preliminary investigations are outlined. The numerical study of the HFMI treatment for two different geometries is reported. Therewith, the influence of various aspects of the simulation process on the modelled RS are investigated, analogously to the case of WRS in the previous chapter. Both numerical approaches and practical aspects are discussed. Based on the present results, a review of the recommendations from previous studies is made as well. Explicit conclusions for the case of HFMI modelling and recommendations regarding future work are highlighted at the end of this chapter.

Finally, as specific conclusions and recommendations regarding future work over welding and HFMI simulation are presented in the previous respective chapters, a general discussion regarding the present dissertation and a proposal regarding the implementation of the present method in a holistic numerical approach regarding fatigue of metals are presented in chapters 6 and 7.

2 Theoretical Background

2.1 Numerical Investigations

2.1.1 FE Simulation of Fusion Welding Residual Stresses

Ever since De Bernandos patented the first arc weld application in 1887 [27], several metal arc welding types have been developed [170]. Nevertheless, in most cases the same principle is applied: an electric arc, i.e. a flow of ions between an electrode and the metallic part or parts, which are to be welded, is established leading to rapid increase of temperature and surpassing the melting point of the welded material (see [97], [179]). Through cooling-down and resolidification of the molten material, the desired connection is achieved. During the investigation of the welding process, a multi-physics problem is arising, as thermal, microstructural, mechanical, and electromagnetic phenomena among others, are present. A thorough overview of the physics of welding, provided by Francis et al. in [51], is presented in Figure 2.

Consecutive increase and decrease of temperature causes respectively the expansion and shrinkage of the material in and near the weld. Restraints from neighbouring unaffected material during both heating and cooling in combination with softening of the material due to increased temperatures introduces significant plastic strains in the weld, which are accompanied by residual stresses. Based on simplified models of thermal expansion and shrinkage and material behaviour, theoretical profiles of WRS have been proposed in the past ([96], [139] etc., see Figure 3). Nonetheless, as this thermal treatment can influence significantly the microstructural composition (phase changes, recrystallization etc.) and in extension the material behaviour of the welded area, the real strains and residual stresses are not always in agreement with textbook knowledge. Weld simulation with the FE method is a powerful tool that can tackle this problem, calculating shrinkage and profiles of WRS with satisfying accuracy taking into consideration all the predominant factors.

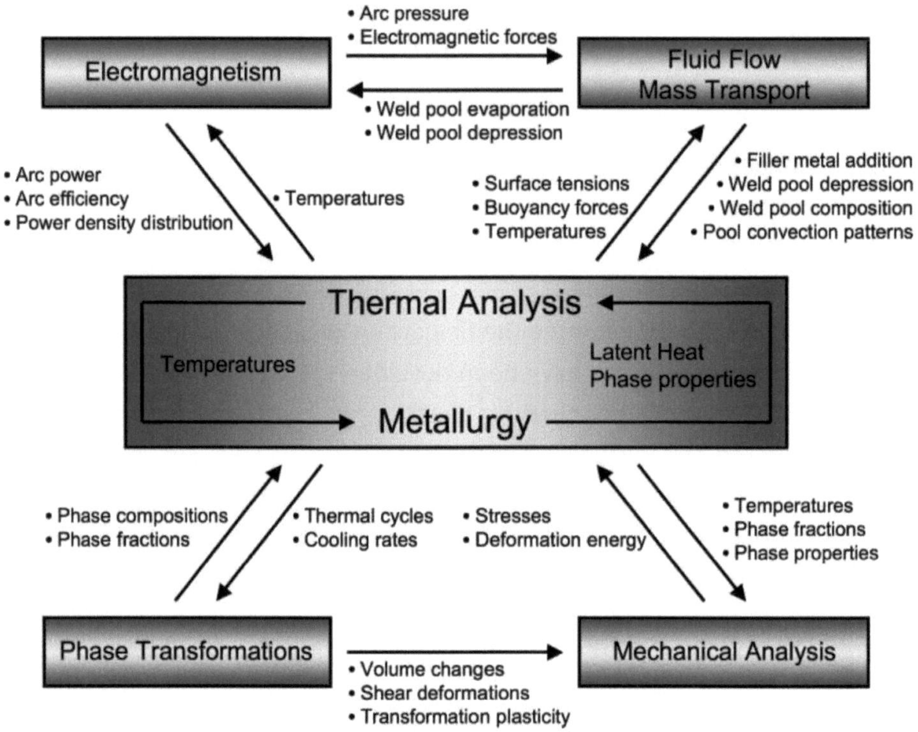

Figure 2: Multiphysics of welding - The processes and respective interactions during arc welding as presented by Francis et al. [51]

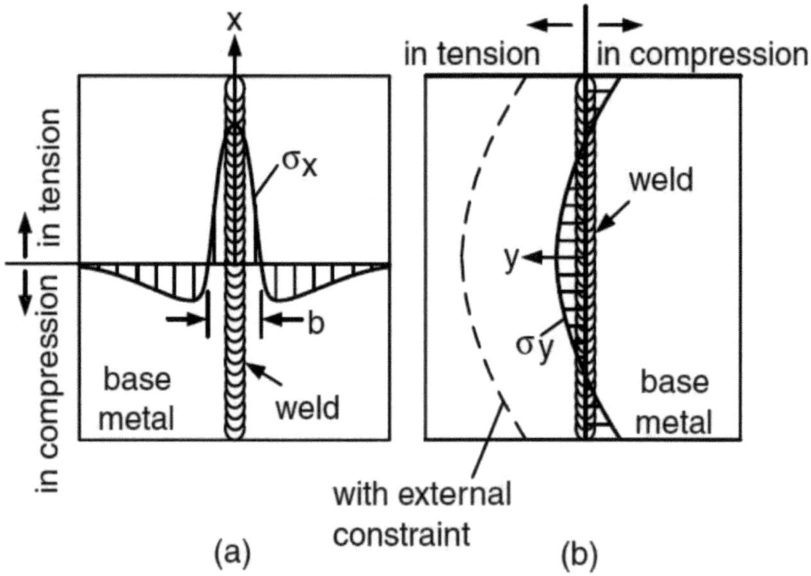

Figure 3: Theoretical profiles of WRS found in [96]: a) Longitudinal WRS; (b) Transverse WRS, "with external constraint" stands for the case of clamped longitudinal sides

The field of computational welding mechanics has widely expanded over the last decades. Although the first finite element formulations regarding coupled thermomechanical simulations of welding were published in the '60s, with the investigations by Argyris et al. being the earliest known to the author of the present dissertation [5], it was not until the '80s when the conditions for the evolution of practical weld simulation were met. In 1984, Goldak presented his approach for modelling of the welding heat source [61], which is until now considered state of the art, while equations for prediction of microstructural transformations were either already available [94] etc., or were proposed around the same time [104]. Furthermore, sufficient computational capabilities became available. Ever since, different approaches for simulation of welds have been proposed, using FE software either of general purpose ([1], [11], [92] etc.) or weld-specialized ([21], [69], [71] etc.).

In many cases, calculation of WRS and welding distortion has been the main subject of the proposed models. For such an engineering analysis though, it is not required to take into consideration all the above-mentioned phenomena. The thermal, the microstructural and the mechanical fields have to be modelled and electromagnetic, fluid flow and mass transfer effects can be neglected (see [92], [111], [139]). Thermal transfer is predominant, as it influences the welding residual stresses both directly by inducing thermal strains and indirectly by influencing the microstructural transformations and in extension the mechanical behaviour of the welded component. These microstructural transformations cannot be neglected in most cases, as they can cause significant discrepancies to the material behaviour during a thermal cycle. For instance, yielding behaviour of the material, whose influence on the order of magnitude of the WRS is predominant, can radically change in the HAZ and the FZ of aluminium welds due to recrystallization of the microstructure [93]. Finally, the mechanical field incorporates the above-mentioned effects in order to calculate the requested output.

These three fields of weld simulation interact with each other as it is presented in Figure 4. For example, thermal behaviour influences directly the phase proportions of ferritic steels, but each phase has its unique thermal properties and as a result, thermal behaviour changes inside a thermal cycle. Moreover, heat transfer material properties are influenced from the microstructure and deformation can cause exchange of heat energy [139]. At the same time, mechanical loading of the heated areas of a component from neighbouring unheated material or applied restraints can influence microstructural behaviour, as in the case of Transformation-Induced Plasticity (TRIP), which is described thoroughly in [42] and [67]. Nonetheless, the influence of these phenomena on calculated WRS is negligible and therefore, the backwards influence of mechanical field on thermal and microstructural field and of microstructural on thermal are usually neglected (see for instance [92], [111] and [139]). Sole exception is the TRIP, whose influence on the WRS is non-negligible and consequently has to be taken into consideration [87], [92].

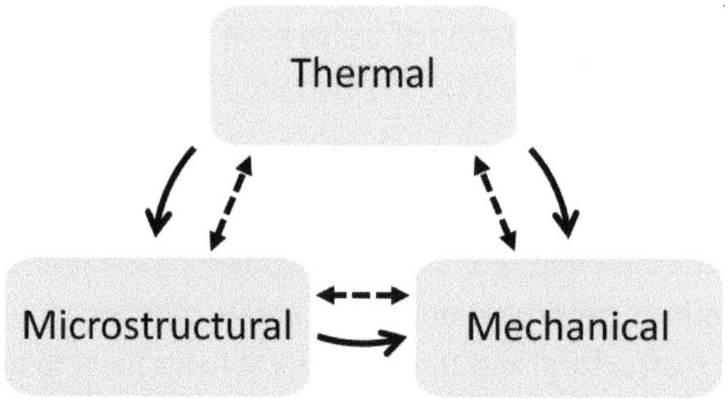

Figure 4: Investigated fields and respective interactions in an engineering approach for arc welding simulation – Arrows with broken and continuous contour are symbolizing the existing and the considered interactions respectively

2.1.1.1 Modelling of Thermal Field

As already mentioned, modelling of the thermal field includes the solution of the heat-transfer problem. Ensuring sufficient preciseness during the calculation of the WRS requires accurate modelling of the heat transfer inside the component and use of temperature-dependent material properties. As mentioned above, the influence of microstructure on thermal material behaviour is neglected and thermal properties of parent material are usually applied for the whole investigated thermal cycle [3], [21], [92]. Loading and boundary conditions of the problem are the heat input from the weld heat source to the weld pool and heat losses from the component to the environment respectively.

Welding Heat Source

The double ellipsoidal model of Figure 5, which was proposed by Goldak et al. in 1984 [61] is still considered state of the art for modelling the heat input into the weld pool during simulations of metal arc welds (see [28], [71], [92] etc.). It can be properly modified for different types of welds as well, such as in [128] for laser welding. The model describes the power density distribution of each point around the centre of a moving heat source as a function of time, and position of the point about the centre, by using quadrants from two different ellipsoidal sources. The two quadrants have common width and depth but different length in order to properly model the elongation of the power heat distribution rear to the centre, due to the movement of the heat source. The power density distribution of the rear and front quadrant are described respectively by the following two Equations[1,2,3]:

$$q(x,y,z,t) = \frac{6\cdot\sqrt{3}\cdot f_r\cdot Q}{a\cdot b\cdot c_2\cdot\pi\cdot\sqrt{\pi}}\cdot e^{-3\cdot\frac{x^2}{a^2}}\cdot e^{-3\cdot\frac{y^2}{b^2}}\cdot e^{-3\cdot[z+v\cdot(\tau_{lag}-t)]^2/{c_2}^2}, \tag{1}$$

$$q(x,y,z,t) = \frac{6\cdot\sqrt{3}\cdot f_f\cdot Q}{a\cdot b\cdot c_1\cdot\pi\cdot\sqrt{\pi}}\cdot e^{-3\cdot\frac{x^2}{a^2}}\cdot e^{-3\cdot\frac{y^2}{b^2}}\cdot e^{-3\cdot[z+v\cdot(\tau_{lag}-t)]^2/{c_1}^2}. \tag{2}$$

The effective heat input rate Q is calculated as follows:

$$Q = \eta\cdot V\cdot I. \tag{3}$$

Proposed values for the weld metal arc efficiency η are given by Dupont and Mader in [32]. These values can differ slightly from those applied by welding engineers in practice (± 5 %), as it is discussed in [57]. It was shown in [58] that this discrepancy causes negligible differentiation of the calculated WRS, which lies inside the acceptable error for practical weld simulation (± 10 % to the calculated WRS).

---

[1] a list of notations is given at the begining of the present dissertation at page xxxv

[2] front quadrant in Figure 5 is the one with dimensions *a, b,* c_1 (Equation (2))

[3] f_r and f_f of Eq. (1) and (2) are according to [186] the fractions of heat deposited in the rear and front quadrant respectively, with $f_r + f_f = 2$

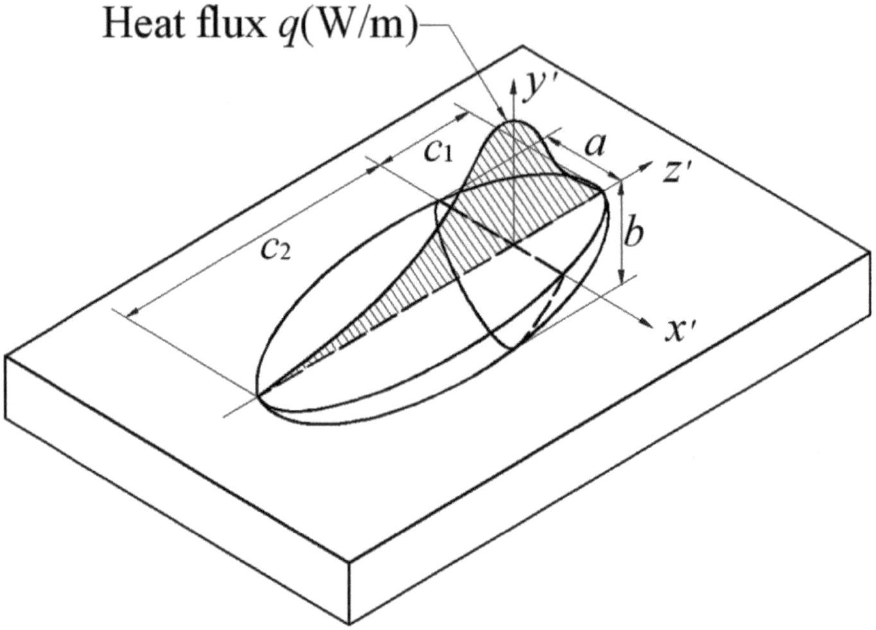

Figure 5: The Goldak's heat source model, figure found in [186]

Heat Transfer

Heat transfer inside the component is governed by the following transient Equation (4):

$$\rho \cdot c \cdot \frac{\partial T}{\partial t} = \dddot{q} + \frac{\partial}{\partial x}\left(K_x \cdot \frac{\partial T}{\partial x}\right) + \frac{\partial}{\partial y}\left(K_y \cdot \frac{\partial T}{\partial y}\right) + \frac{\partial}{\partial z}\left(K_z \cdot \frac{\partial T}{\partial z}\right), \tag{4}$$

derived by combining the 1st thermodynamics law,

$$\Delta U = Q - W, \tag{5}$$

which is a version of the law of conservation of energy adapted for thermodynamic systems, with Fourier's law of heat conduction,

$$q = -K\nabla T, \tag{6}$$

and neglecting the influence of fluid flow inside the weld pool on heat transfer.

Heat Losses

Heat losses from the component to the environment through convection and radiation constitute the boundary conditions of the thermal problem. Heat loss due to convection and radiation are described by Newton's law of cooling,

$$\frac{q}{A} = h_c \cdot (T_s - T_b), \tag{7}$$

and Stefan-Boltzmann's law,

$$\frac{q}{A} = h_r \cdot (T_s - T_b)^4, \tag{8}$$

respectively. During the first weld-modelling applications, a temperature independent value for the coefficient of heat losses was applied [3], [5]. In the latter case, a more delicate approach was followed by applying a coefficient for the material in the FZ, accounting for radiation effects as well. In any case, the influence of radiation on heat losses at room temperature is negligible, but cannot be neglected as it is becoming predominant at high temperatures. Therefore, its modelling is necessary for precise thermal results inside and near the weld pool. Newton's law of cooling, albeit is not anymore physically valid when radiation becomes predominant, was applied with a modified coefficient for heat losses, in order to account for the radiation effects as well in [5]. This approach has ever since provided sufficient results ([1], [44], [140], [92] e.g.). Proposed values for the total coefficient of heat losses h_T applied instead of h_C, were calculated in [1] based on data found in [18].

2.1.1.2 Modelling of Microstructural Effects

Some microstructural properties of the metallic alloys influence their macroscopic mechanical behaviour. Their differentiation due to rapid heating and cooling inside a welding thermal cycle has to be taken into consideration during welding simulation, in order to ensure that WRS are calculated with required preciseness.

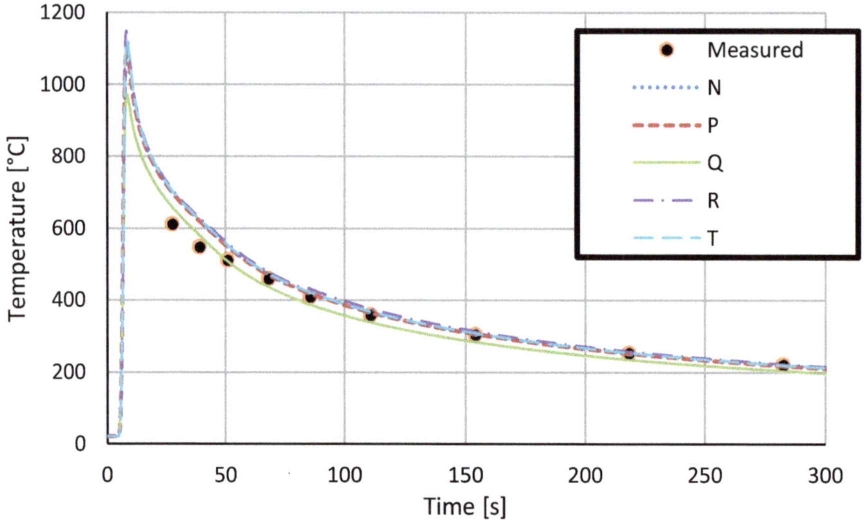

Figure 6: Sensitivity analysis for the heat source parameters (N to T are different sets of heat source parameters a, c, f_r and f_f) and real temperature measurements of a thermal cycle in the HAZ during welding for HT36 steel – Published in [92], measurements and welding parameters were found in [3]

Microstructural Phases of Steel

Steel is a multi-phase solid, where each phase represents a different crystal system of the Fe and C atoms [67]. The prevailing phase, for a plain Fe-C alloy is dependent on temperature and mass percentage of C, as it is displayed at the phase diagram of Figure 7. Each phase demonstrates different mechanical behaviour. In alloys, which contain more elements, the phase composition is as well influenced by their percentages.

Austenitic transformation[4] begins and ends during heating, when austenitization temperatures *A1* and *A3* respectively are reached. The following Equations, which are found in [60] and [111], were proposed respectively for the calculation of both temperatures, based on chemical composition of the investigated steel alloy:

$$A1 = 723 - 10.7 \cdot Mn - 16.9 \cdot Ni + 29 \cdot Si + 16.9 \cdot Cr + 290 \cdot As + 6.4W, \tag{9}$$

[4]austenitic transformation is the transformation of another steel phase to austenite - the reversed transformations of austenite to ferrite (ferritic transformation), bainite (bainitic transformation), pearlite (pearlitic transformation) and the rest of the phases apart from martensite (martensitic transformation) are conventionally called austenitic as well (see [104])

$$A3 = 912 - 200 \cdot \sqrt{C} - 15.2 + 44.7Si + 315 \cdot Mo + 13.1 \cdot W - (30 \cdot Mn + 11 \cdot Cr + 20 \cdot Cu - 700 \cdot P - 400 \cdot Al - 120 \cdot As - 400 \cdot Ti). \tag{10}$$

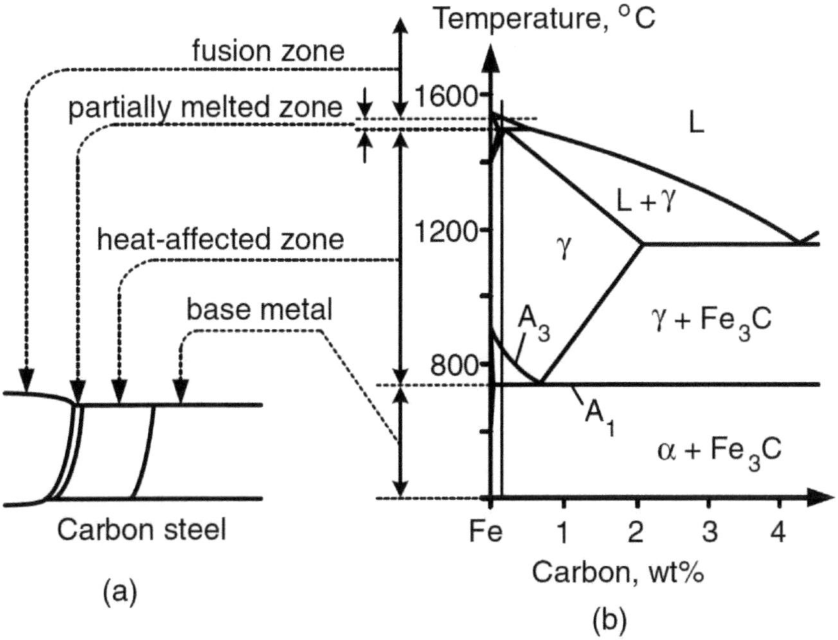

Figure 7: Carbon steel weld: a) HAZ; b) Phase diagram (found in [96], γ stands for austenite and α for ferrite)

Nevertheless, the information of the Fe-C diagram and the Equations (9) and (10) are referring to "static" theoretical values (*dT/dt* → 0), as they are obtained by respective experiments. During dynamic heating cases like welding, the boundaries of the phase changes are being shifted. Leblond and Devaux observed that when dynamic heating takes place, a deviation between real and theoretical values of *A1* and *A3* temperatures is met, with the real values being higher (see [104]). In the case of *A1* though, the increase could be considered negligible. This effect was named "retardation" by Leblond and Devaux and was further investigated in [116] and [133]. Macedo et al. correlated the real values of *A3'*, hereinafter referred to as "dynamic", with the heating rate for the investigated steel and *A1*, as it is presented in Equation (11),

$$A3' = \omega \cdot \left(\frac{dT}{dt}\right)^{-\varphi+1} + A1, \tag{11}$$

where ω and φ are constants with proposed values of 135.6 and 0.95 [116]. Applying $A3'$, decouples the influence of T_{max} and t_a, which describes the transient nature of austenitic transformation. The dynamic effect is incorporated in the calculation of $A3'$. Hence, the influence of t_a for single-pass welds (single thermal cycle) can be neglected.

The Fe-C phase diagram, although being inadequate to describe precisely the dynamic nature of microstructural transformations during welding, can act qualitatively and advisory. At an initial stage of a weld analysis, it can offer a first rough evaluation of the predominant microstructural transformations in the different parts of the weld, based on the locally reached maximum temperature (T_{max}) during heating. A thorough overview of the zones with different microstructure in and near the weld pool based on the phase diagram of FE-C is given in Figure 7. In the fusion zone (FZ), when the temperature exceeds the melting point (T_{melt}), full austenitization at first and full melting later take place. Material in the HAZ transforms partially to austenite, away from the FZ, or even completely, closest the boundaries with the FZ, without exceeding the fusion temperature. Between the FZ and the HAZ, lies the thin zone of material, where partial melting takes place. The microstructure of the "parent" or "base" material (PM of BM) outside the HAZ, remains uninfluenced by the heat treatment.

During cooling, the proportion of the austenitized material transforms again. However, the resulting microstructure deviates from the respective one in the phase diagram of the parent material as rapid cooling takes place after welding. Final microstructure after cooling can differ significantly depending on the cooling rate. Conventionally, the cooling time from 800 °C to 500 °C (t_{85}) is used as index for the cooling rate. Rapid cooling rates (t_{85} < 10 s) lead to the creation of martensitic microstructure (martensitic transformation). Nevertheless, due to the rapid change, full transformation may not be achieved and remaining austenite can be met in the final microstructure. If lower cooling rates take place and cooling time is in the order of magnitude of 50 s, austenite transforms to pearlite instead (pearlitic transformation). For slow cooling rates ($t_{85} \approx$ 100 s), austenite transforms to bainite or ferrite (bainitic of ferritic transformation). For each alloy the exact boundaries between the various possible transformations may deviate from the above rough description due to the above-mentioned dependence of phase transformations on chemical composition. Continuous cooling transformation (CCT) diagrams (see [147], [148], [188]) as the one presented in Figure 8, illustrate the influence of cooling rate on the microstructural transformations, depending on the chemical composition of the parent material.

State of the Art Models for the Simulation of Phase Changes

Austenitic transformation, whose influence on WRS is discussed in [177], takes place through diffusion [67], while martensitic transformation is a non-diffusional thermal-driven process and takes place through rearrangement of all the atoms of the crystal lattice [67], [111]. The different nature of each of the two transformations dictates the use of different prediction models. The models proposed by Leblond and Devaux in [8] and Koistinen and Marburger in [9] were widely applied in the last decades (see [74], [112], [184] etc.) for modelling austenitic and martensitic transformations respectively. The semi-empirical JMAK model developed by Johnson-Mehl [85], Avrami [7] and Kolmogorov [95] and the Kirkaldy model, which is described in [60], have been as well applied as an alternative to the Leblond model, for considering the several phase transformations of various metals (see for instance [8], [9] and [154]).

The Leblond-Devaux model was formatted initially for two existing phases and a unique transformation and was then generalized for the case of n phases and several possible transformations, for both heating and cooling. The evolution of microstructure n is governed by Equation (12)

$$\dot{n}_i = \frac{{n_i}^{eq}(T) - n_i}{\theta(T)}. \tag{12}$$

The model is based on the existence of an equivalent volume fraction of the transformed phase n_{eq}, which is different at each temperature level. The model is based on a rational assumption: if enough time is provided, the microstructural transformations will end when this equivalent fraction is reached. In the case of austenitic transformation during heating, n_{eq} for the austenitic phase is equal to 0 and 1 at *A1* and *A3* temperature respectively. Yet, calibration for each investigated material is required, as $\vartheta(T)$ is dependent on chemical composition as well [104].

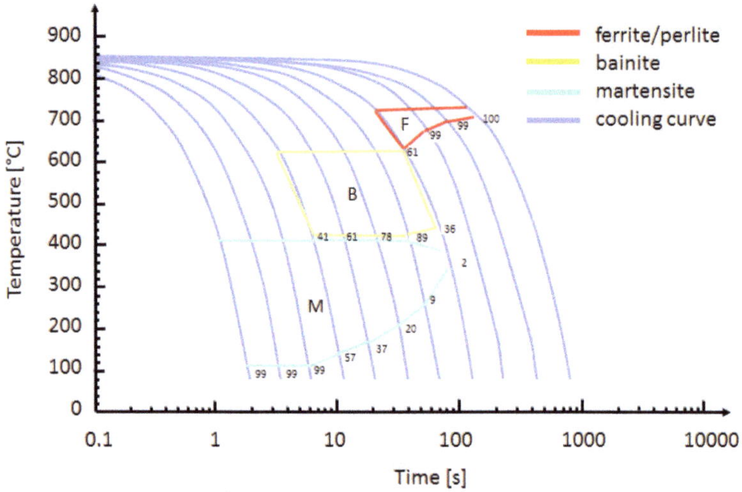

Figure 8: CCT diagram of steel S355, found elsewhere [69][5]

On the other hand, Koistinen and Marburger proposed a straightforward model [94], correlating the fraction of the remaining austenite after the martensitic transformation with the transformation starting and the quenching temperature (final cooling temperature), as it can be seen in following Equation (13)

$$n_i^{aust} = e^{(-1.10 \cdot 0.01(M_S - T_q)}; \; with \, M_s > T_q > -80 \text{ °C}. \tag{13}$$

Martensite starting temperature M_s, can be either calculated analytically by Equation (14)

$$M_s = 561 - 474 \cdot C - 35 \cdot Mn - 17 \cdot Ni - 17 \cdot Cr - 21 \cdot Mo, \tag{14}$$

which is found elsewhere [60], [111], or taken directly by the respective CCT diagram.

More recent work has shown that the mechanical behaviour of a component can be sufficiently described by major parameters of the thermal cycles. The STAAZ model presented by Ossenbrick and Michailov in [134] completely overtook the need for transient microstructure calculation by carrying out mechanical behav-

[5] the diagram was redrawn by the author of the present dissertation in order to improve the figure quality

iour tests on higher temperature levels and strain rates, to characterize the material behaviour of the different component areas (i.e. areas with different thermal cycles) during a single-pass weld. A triple parameter consisting of T_{max}, t_{85}, and t_a was proven to be sufficient to describe the material behaviour inside a thermal cycle. This statement is becoming easily understood, based on the above-described influence of these three parameters. Mechanical material behaviour of the investigated material inside a thermal cycle for different values of this triple parameter were experimentally acquired. A finite element analysis was subsequently carried out, wherein each investigated component was divided to different areas, as in the present approach. During simulation, each area should then be assigned with a respective, experimentally acquired material behaviour depending on its triple parameter values. In this way, the material behaviour during welding could be adequately modelled. Experimental results from later work have shown that the method could be extendable to multi-pass welds [19] as well. In order to apply the STAAZ model though, there is a need to repeat the delicate calibration tests for every newly investigated material.

The Gkatzogiannis – Knoedel – Ummenhofer Engineering Approach

Alternatively to the above-mentioned methods, an engineering approach was proposed for the simulation of WRS in a study, which was carried out in the framework of the present doctoral dissertation [92]. The proposed method is in its original form predictive, calculating the WRS based on existing CCT diagrams. Theoretical *A1* and *A3* temperatures were calculated based on Equations (9) and (10) respectively. The dynamic *A3'* temperature taking into consideration the retardation effect was calculated based on Equation (11). During heating, the whole component was simulated with parent material behaviour. When T_{max} was between *A1* and *A3'* a linear interpolation between no and complete austenitization was carried out, in order to calculate the percentage of austenitized fraction (see Figure 9). During cooling, the austenitized behaviour was assumed to behave according to the respective CCT diagram. Several material models[6] for different levels of austenitization during heating and cooling paths were simulated. Each material model was characterized by its T_{max} and t_{85}. Applying *A3'* decouples the

[6] hereinafter called Cooling Material Model (CMM)

influence of T_{max} and t_a, which describes the transient nature of austenitic transformation. The dynamic effect is incorporated in the calculation of *A3'*. Hence, the influence of t_a for single-pass welds (single thermal cycle) can be neglected and only a double parameter consisting of T_{max} and t_{85} is applied. Each FE-element in the HAZ and the FZ was assigned during cooling with the closest material behaviour, different from that of the parent material, based on the double-parameter (T_{max} and t_{85}, see Figure 10). The double-parameter of each finite element of the model was calculated during the solution of the thermal field [92]. Preciseness of the method, regarding the consideration of microstructural influence, depends on the number of simulated material models.

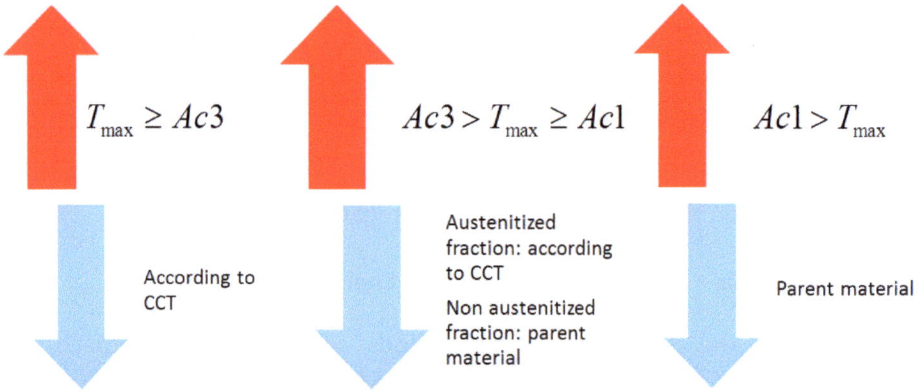

Figure 9: Modelling principle of heating and cooling in the Gkatzogiannis-Knoedel-Ummenhofer engineering approach, *A1* and *A3* are symbolized as *Ac1* and *Ac3* respectively, figure found in [69]

At each temperature level, the yield strength[7] was calculated by a linear mixtures law (Equation (15)), which is widely applied for calculating the material parameters of multi-phase solid materials (m phases) ([60], [111] etc.):

$$\sigma_Y = \sum_{i=1}^{m}\{(n_i \cdot \sigma_{Y_i})\}. \tag{15}$$

[7] along the manuscript the symbol σ_y is applied for yield strength, although in literature the symbol f_y can be met as well

Apart from the phase fractions, influence of austenite grain size and transformation induced plasticity[8] (TRIP) were investigated in [69], as theoretically they can have a significant effect on the calculated WRS (see [60], [111], [71] etc.).

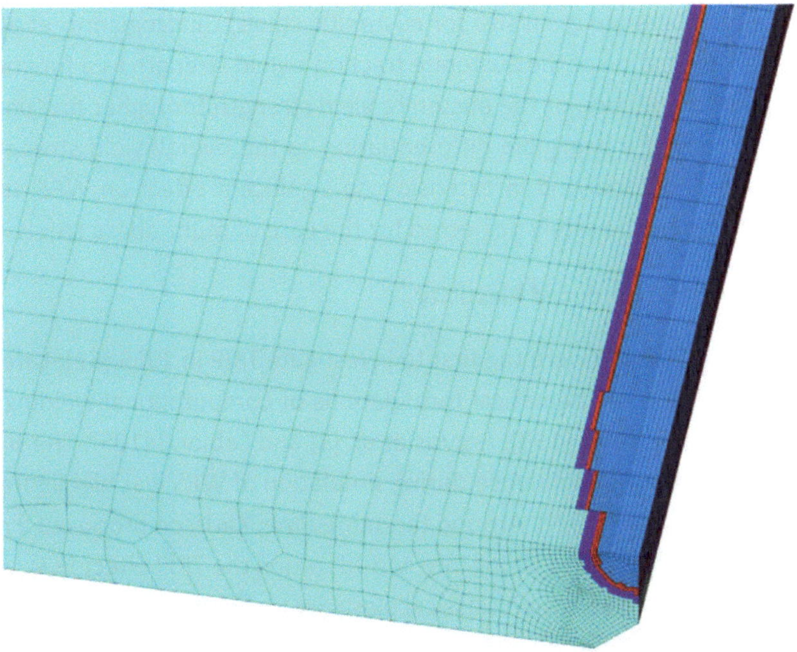

Figure 10: Application example of the proposed method: the areas of the component, with different material models during cooling-down: CMM4 was assigned to blue areas; CMM3 was assigned to red areas; CMM2 was assigned to purple areas; rest of the plate is preserving CMM1 (parent material)

Austenite Grain Size

Austenite grain size directly influences the yield strength of the austenitic phase. The austenite grain growth was analytically calculated in [92] through equations found in literature. Considering that T_{max} is the dominant temperature during the austenitization time, therefore assuming $T = T_{max}$ during the austenitization time, Equation (16) by Lee and Lee [107] was applied, in order to calculate the austenite grain size as a function of austenitization time, temperature and chemical composition, for low alloy steels:

[8] Transformation Induced Plasticity is defined by Fischer et al. [42] as follows "... significantly increased plasticity during a phase change. For an externally applied load for which the corresponding equivalent stress is small compared to the normal yield stress of the material, plastic deformation occurs ..."

$$d_{aust} = 76671 \cdot e^{\left(-\frac{-89098+3581\cdot C+1211\cdot Ni+1443\cdot Cr+4031\cdot Mo}{R\cdot T}\right)} \cdot {t_a}^{0.211}. \tag{16}$$

Knowing the austenite grain size and using Norström's [131] correlation of yield strength to the austenite grain size for austenitic steels, Gkatzogiannis et al. derived a rough estimation of the austenite fraction's yield strength as a function of t_a [92]. The differentiation of the austenite yield strength for higher austenitization times produced a negligible differentiation to the overall yield strength of the investigated material and in extension no significant effect on WRS. It was concluded that it can be safely neglected for the case of multi-phase steels.

TRIP: Transformation Induced Plasticity

Greenwood and Johnson were the first to investigate TRIP in 1965 [62]. Later on, TRIP was included in mathematical formulations of computational welding mechanics by Rammerstorfer [138], Argyris et al. [5] and Leblond [105]. The significance of considering TRIP influence during simulation of WRS was highlighted by Oddy et al. in 1989 (see [132]). Around the same time, Karlsson proposed a more straightforward approach for taking into consideration the TRIP effect [87], which is applicable even to cases, when classical plasticity theory is applied and no complicated mathematical formulations of material behaviour are required. Karlsson's proposal was the arbitrary lowering of the yield stress in the respective regions of the HAZ during cooling. This approach was adopted as well by Gkatzogiannis et al. in [92], when martensitic transformation took place.

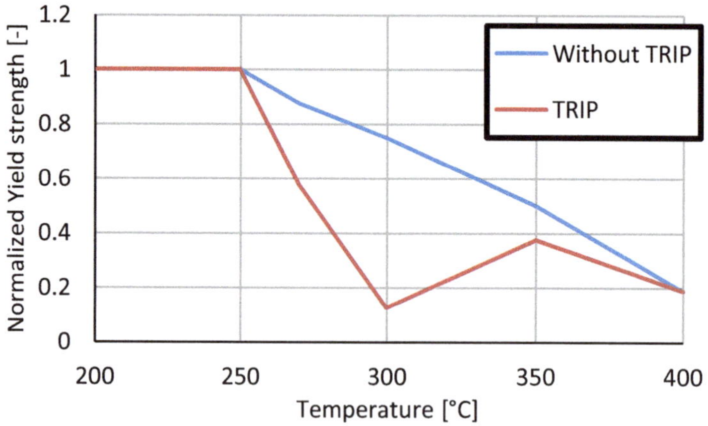

Figure 11: The arbitrary reduction of yield strength in the respective temperature range proposed by Karlsson for the consideration of TRIP during welding simulation, based on a diagram from [87]

2.1.1.3 Modelling of Mechanical Behaviour

Analysis Type

Inertia effects during welding are negligible but plastic strains are introduced; hence, a quasi-static analysis for non-linear material is carried out for modelling the component's structural behaviour. The quasi-static analysis for a system of finite elements is governed by the following Equation[9, 10]:

$$\mathbf{f}_s(\mathbf{u}) = \mathbf{p}(t). \tag{17}$$

In the case of structural analysis during welding simulation, the thermal strains caused by the welding thermal cycle constitute the imposed loading. They can be calculated with the following Equation:

$$\varepsilon^{th} = a^{se}(T) \cdot (T - T_{ref}), \tag{18}$$

based on the FE nodal thermal history, which is previously calculated during the thermal transient analysis, and they are applied as structural nodal loads (see [92]).

Dilatometry

The coefficient of thermal expansion α (secant or instantaneous) is strongly dependent on microstructure. According to common engineering knowledge, ferritic and austenitic steels have a coefficient of thermal expansion equal to $12 \cdot 10^{-6}$ and $16 \cdot 10^{-6}$ respectively at room temperature, both variating slightly depending on the specific microstructure of each investigated alloy and the temperature. All phases of steel apart from austenite are usually assumed to have the same dilatation behaviour as ferrite. When the microstructure is multiphase (m phases), α can be calculated by applying once again a linear mixtures law,

---

[9] use of bold letters states tensor notation

[10] due to the inelastic behaviour of the material the force-displacement relation is path-dependent, thus the resisting force vector is an implicit function of displacement $\boldsymbol{f}_s = \boldsymbol{f}_s(\boldsymbol{u})$ and no more equal to $\boldsymbol{ku}$, where $\boldsymbol{k}$ is the stiffness matrix, as in the case of elastic material behaviour

$$a^{tot}(T) = \sum_{i=1}^{m} \{n_i(T) \cdot \alpha^{ni}(T)\}, \tag{19}$$

with such an approach having been adopted elsewhere (see [39], [92]). The dilatometric diagrams, such the one given by Ferro and Bonolo [39] (see Figure 12), provide a thorough overview over the microstructural and temperature dependency of the dilatation behaviour. The instantaneous coefficient of thermal expansion, which is equal to the slope of the dilatometric curve, changes significantly during heating and cooling after the completion of the respective microstructural transformations. During the transformations a significant jump in the diagrams is met, which corresponds to the volumetric changes caused by the rearrangement of the atoms in the crystal lattice, during changes from α-phase (ferrite) to γ-phase (austenite) and vice versa. After completion of the phase changes a significantly different slope of the dilatometric curve is observed, during both heating and cooling.

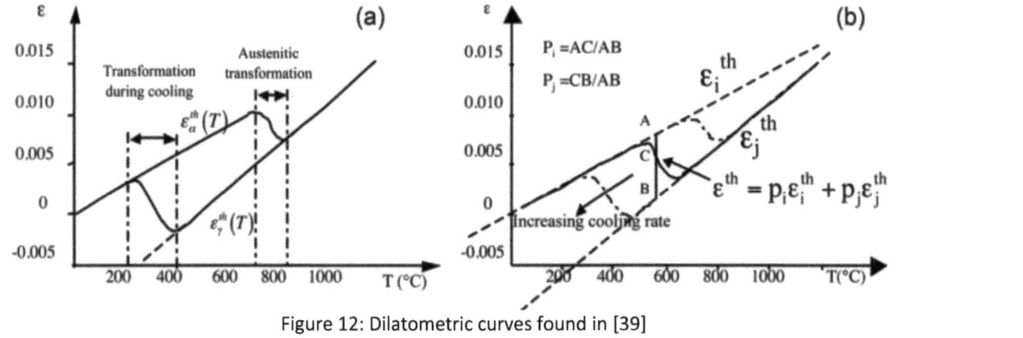

Figure 12: Dilatometric curves found in [39]

Furthermore, as can be seen in Figure 12b, differentiated cooling rate influences the dilatation behaviour as well. As mentioned above, changing cooling rate leads to different phase-transformations taking place, which have different starting temperatures. Therefore, transition from austenitic dilation behaviour to that of another phase, is taking place at a different temperature range. Increasing of cooling rate causes the decrease of the transition temperature and vice versa. This is in accordance with the general form of the CCT diagrams as the one in Figure 8. Higher cooling rates lead to martensitic transformation, which takes place at lower temperatures (left side of the CCT diagram). Independently of the cooling rate and in extension of the final microstructure after cooling, the behaviour of the material shifts back to the dilation curve of ferrite. Therewith, the above-mentioned assumption that all phases apart from austenite display same dilation behaviour with ferrite is confirmed.

Albeit the ending point of the dilation curve is identical independently of this transition temperature range during cooling, the influence of cooling rate on WRS is significant. When the increase of thermal strains due to transition from austenitic to ferritic dilation behaviour takes place in a lower temperature range, it is accompanied by a higher yield strength, leading in this way into the creation of higher plastic strains and in extension higher WRS. The above-statement is valid, only when the influence of microstructure on yield strength is negligible. Nevertheless, as the yield limit is for many alloys not only temperature-dependent but microstructure-dependent also, as it was described above, the interpretation of the influence of cooling and dilation is becoming even more complex. Understanding this complexity, further confirms the necessity of FE modelling of welding shrinkage and WRS. A direct modelling of the dilation behaviour of the investigated material for heating or a specific cooling rate through direct input of thermal strains is more straightforward than the use of the coefficient of thermal expansion based on Equation (19). Gkatzogiannis et al. applied Equation (19) in [92], in order to predict the dilation curves of their CMMs, based on the predicted microstructural transformations. Thermal strains as a function of temperature were then applied as input for the simulation.

Rate-Independent Nonlinear Material Behaviour

In order to calculate the plastic strains and the WRS caused by the external loading, plasticity of the investigated material must be considered. The non-linear material behaviour of steel can be simulated as bilinear, multi-linear, with Ramberg-Osgood model [137] etc. However, as the maximal plastic strains during welding are usually lower than 2 % [92], a bilinear model with a tangent modulus *H* (slope of the second branch), which is selected based on the stress for this introduced plastic strain, can be considered sufficient for welding simulation [55]. The von-Mises yield criterion,

$$f(\sigma) = J_2 - \tau^2, \tag{20}$$

which is widely applied for simulating the yield behaviour for metallic components, is applied as well during simulation of welding. Temperature-dependent material parameters are usually taken into consideration (see [3], [21], [28], [92] etc.).

Zhu and Chao have nonetheless proven in [190] that an engineering approach, whereby only the predominant yield strength is considered to be temperature-dependent, can offer results with sufficient preciseness. Moreover, even in cases where more material properties were temperature-dependent, usually a threshold temperature was defined, above which either stiffness was set to zero [164] or effects of thermal treatment were neglected [165], [52]. This threshold temperature was named *"cut-off"* by Lindgren (T_{cut}). It varies from 500 °C up to 900 °C in [52], [164] and [165]. Preciseness was not compromised in these cases. The influence of defining and variating T_{cut}, from up to T_{melt} down to 600 °C, was investigated by Tekriwal and Mazumber in [160]. It was concluded that at lowest T_{cut} a maximal overestimation of 15 % of the WRS took place. The principal of T_{cut} is thoroughly described by Goldak and Akhlaghi in [60] and Lindgren in [111]. In any case, it has to be considered that attaining material properties at higher temperatures (>1000 °C) is subject to significant testing restrictions.

Rate-Dependent Nonlinear Material Behaviour

Lindgren [111] proposed that during welding simulation, when very high accuracy is required, strain rate dependent plasticity should be considered, without providing though any information regarding the order of magnitude of the influence on calculated WRS. Strain rate was introduced quite early into the formulation of finite element equations for modelling the thermomechanical welding process. Argyris et al. proposed in 1984 a viscoplastic material model, wherewith the strain rate dependency of the yield behaviour[11] is taken into account [5]. However, prior to an investigation from Gkatzogiannis et al. [59], no quantified influence of strain rate dependency on simulated WRS was known to the author of the present study. In this study that was carried out in the framework of the present doctoral dissertation, the effect of strain rate dependency on WRS was investigated and quantified. Preliminary analyses showed that in the heat affected zone (HAZ) and

[11] hereinafter called simply strain rate dependency

the fusion zone (FZ) of a 3-pass butt-weld, strain rates of up to 0.122 s^{-1} are present. Although this value lies clearly lower than the classical dynamic cases such as modelling of ballistic tests ($\dot{\varepsilon} \approx 10^6\, s^{-1}$) or car crash simulation ($\dot{\varepsilon} \approx 100\, s^{-1}$), still clearly deviates from the static case ($\dot{\varepsilon} \rightarrow 0\, s^{-1}$). The authors applied the material model presented by Perzyna in 1966 in [136]:

$$\dot{\tilde{\varepsilon}}_{pl} = \alpha \cdot \left(\frac{\sigma_y{}'}{\sigma_y} - 1\right)^{\frac{1}{\beta}}, \tag{21}$$

which enables the consideration of strain rate dependency during implicit quasi-static FE simulations. It is analogous to the Cowper-Symonds model [25], as it was presented by Jones in [86]:

$$\dot{\varepsilon}_{pl} = \Delta \cdot \left(\frac{\sigma_y{}'}{\sigma_y} - 1\right)^{\gamma}. \tag{22}$$

It is obvious that for the uniaxial loading case, the two models are becoming equivalent as $\dot{\tilde{\varepsilon}} = \dot{\varepsilon}_{pl}$ and consequently, the coefficients α, $1/\beta$ and γ, δ become respectively equal. Table 1, which was found in [86], provides values for the Cowper-Symonds model, which were gathered from different sources.

Table 1: Coefficients for Cowper-Symonds model, presented in [86]

Material	Δ [s^{-1}]	γ [-]	Reference
Mild steel	40.4	5	Cowper and Symonds [25]
Aluminium alloy	6500	4	Bodner and Symonds [13]
a-Titanium (Ti50A)	120	9	Symonds and Chon [156]
Stainless steel 304	100	10	Forrestal and Sagartrz [50]
High tensile steel	3200	5	Paik and Chung [135]

Hardening Behaviour

The influence of welding thermal cycles complicates the selection of appropriate hardening behaviour as well. The zone along and near the welding line is compressed and subsequently stressed under tension by the subsequent cold material during heating and cooling respectively. Therefore, reversed plasticity takes place, which in the case of multi-pass welding can constitute of more than one hysteresis. Selection of an appropriate hardening model for describing the plastic behaviour after the reversal of loading is predominant for the calculated WRS [111], [151]. Kinematic ([17], [28], [88], [23], [31] etc.), isotropic ([141], [15], [40], [70], [2], [71] etc.) and mixed hardening ([106], [126] etc.) models have been applied in the past for weld simulations. An assumption that the Bauschinger effect of austenitic steels is eliminated during subsequent yielding, when prior yielding takes place at elevated temperatures or the material is heated significantly between consecutive yielding sequences, has been validated by Mataya and Carr in [120]. Hence, even those materials, which exhibit an obvious Bauschinger effect at room temperature, are expected to behave according to isotropic hardening during reversed plasticity. Wohlfart et al. [178], Mullins and Gunnars [126] and Smith et al. [151], [152], who investigated the influence of hardening modelling on WRS simulation of austenitic steels, came to the conclusion that isotropic hardening provides better agreement than kinematic. The latter two proposed as well to use of a mixed hardening behaviour, which was even more accurate. On the contrary, investigations by Gkatzogiannis et al. for ferritic multiphase [55] and austenitic steels [56] have shown a reversed effect, with kinetic hardening providing better agreement than isotropic. Still, the difference between isotropic and kinematic hardening in these investigations was inside the uncertainty boundaries of the applied weld modelling approach. In all above-mentioned comparative investigations though, kinematic hardening under-predicted the WRS in the weld area in comparison with isotropic.

Boundary Conditions

During the structural analysis, appropriate boundary conditions have to be applied, in order to simulate the mechanical restraints of the real component in reality. Components for scientific measurements are usually welded in the laboratory either restraint-free or clamped down. Wherever possible, symmetry conditions can be applied as well in order to reduce the computational time.

According to common practice, modelling of clamping mechanisms is usually done by fixing the displacements of the respective nodes in all directions. Nonetheless, this modelling approach may deviate from reality, whereby clamping mechanisms allow small in-plane displacements to take place. On the other hand, carrying out a restraint-free implicit analysis leads to numerical instabilities (underdetermined static system). These two modelling challenges were discussed by Gkatzogiannis et al. in [55], who proposed the use of linear spring elements for modelling both cases and investigated the effect of boundary conditions on the calculated WRS. Different setups, using either "*soft*" or "*hard*" linear spring elements for modelling of restraint-free and clamped components respectively, were applied. A similar approach for modelling a free plate was applied earlier by Caron et al. in [21] and Gkatzogiannis et al. in [92]. The terms "soft" and "hard" exhibit the empirical nature of the method. In any case, stiffness should be respectively low in the former case in order to allow an almost free movement and high enough in the latter case, so that only very small displacements are possible.

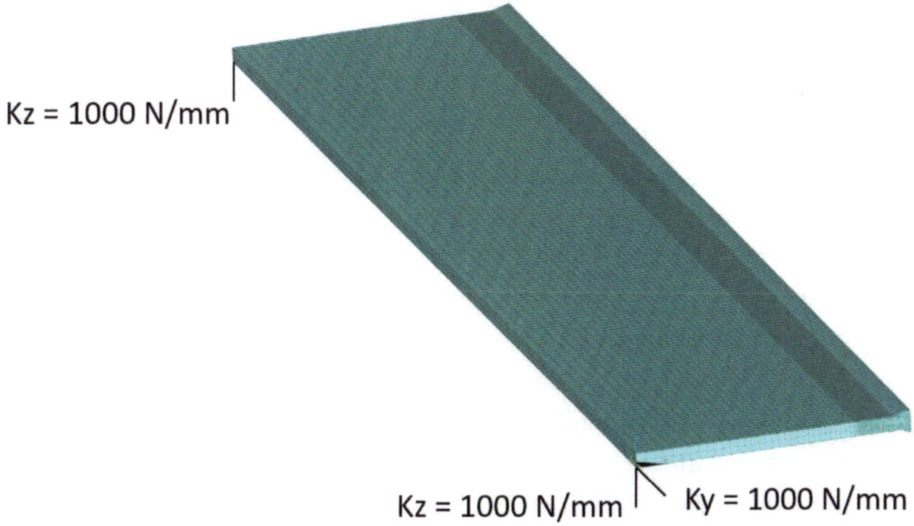

Figure 13: Modelling of restrained-free component by using springs with low stiffness, previously presented in [92]

In the case of the unrestrained plate, two spring elements in the vertical and one in the longitudinal direction were modelled. Displacements to the direction transverse to the welding line were not restrained, as the symmetry conditions at the centreline of the plate constitute already the static system overdetermined in this direction. A stiffness value of 10^3 N/mm for the soft springs was proven low enough in order to simulate with preciseness a restraint-free component [55].

In the case of clamped components, the two set ups of Figure 14 and various values for spring stiffness were tested [55]. The blue region represents the clamped area. In setup (a), spring elements restrained transverse displacements (X-axis) of all nodes at the outer edge of the clamped area. The longitudinal displacement of the upper and lower corner nodes (Y-axis) on both edges of the plate was fixed as well with linear spring elements. Total stiffness was equal in both directions. Vertical (Z-axis) displacement of all the nodes in the area was fixed. In setup (b), spring elements restrained displacements along X (transverse) whilst Y (longitudinal) and Z (vertical) displacement of the blue area were fixed. Values in the range of 10^3 up to 10^9 N/mm for the springs' stiffness were tested for both setups. The results were compared with the case where the nodes of the clamped area were simply fixed to all directions and a significant increase in preciseness was met. Highest preciseness was achieved by applying setup (a) with spring stiffness higher than 10^6 N/mm.

Still, it has to be highlighted that although the improvement in the results is clear, fixing of the required stiffness values and application to other cases is not viable. As above-described, linear springs were applied for restraining the transverse displacement of all nodes along the edge of the clamped area. Therefore, the number of the restrained nodes and in extension the applied spring elements, is dependent on the size of the clamped area, and the mesh density. As a result, the spring stiffness should be re-calibrated in each investigated case. Yet, the results from [55] can be applied as guidance for practical applications based on rough calculations and using the highest tested value of 10^9 N/mm, matching a spring density of approximately 20 x 10^7 (N/mm)/mm in the clamped areas, should lie on the safe side for similar clamping area size and mesh density.

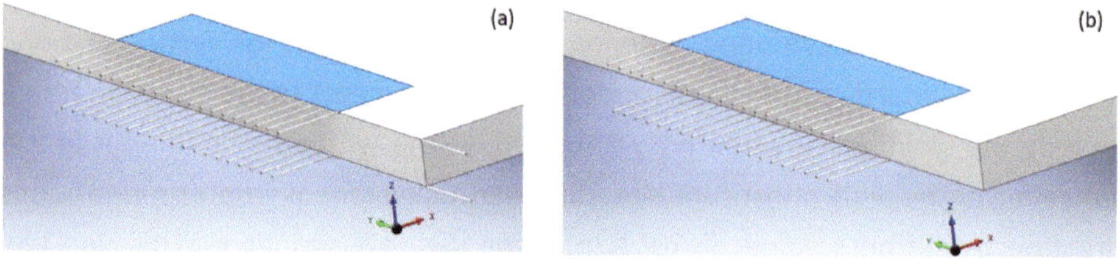

Figure 14: Investigated setups of linear spring elements for modelling the clamping mechanisms [53]

2.1.1.4 Surface Treatment

Previous investigations by Shaw et al. [149] and Hensel et al. [72] have proven that blasting surface treatments, which are applied prior to welding as well, introduce significant RS on the respective components. The following diagram, which was found in [149], displays the magnitude of RS caused by various treatments on the surface of a mechanical component of steel 20MnCr5. It is known to have a nominal yield strength in the order of magnitude of 1150 MPa. If significant stresses are introduced prior to welding from these treatments, this could affect the final calculated WRS profile. No previous numerical investigation of blasting for considering the RS during a welding simulation is known to the author though.

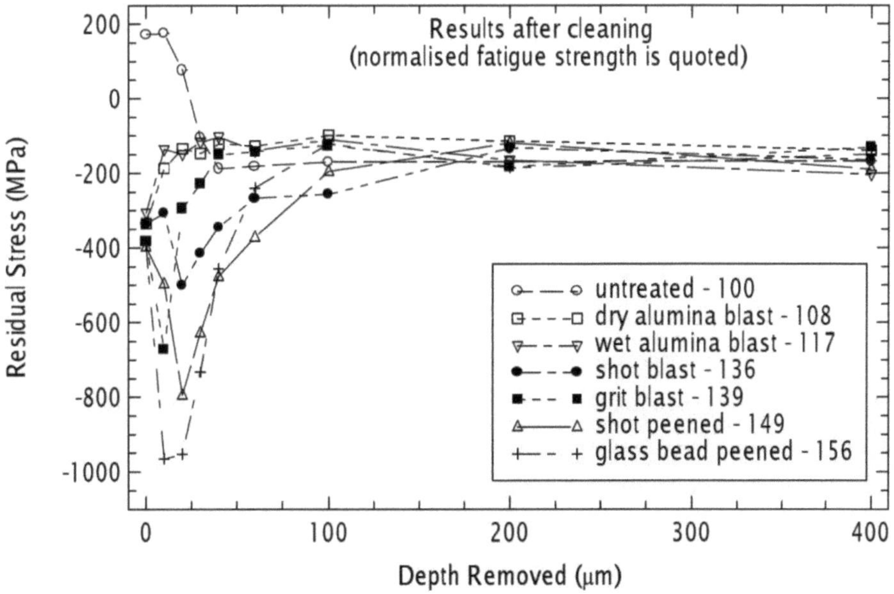

Figure 15: Introduced RS from various treatments on the surface of a mechanical component of steel [149]

2.1.1.5 Specimen Cut

Most of the fatigue testing results are extracted from small specimens, which are cut out of larger welded plates. A relaxation of WRS is supposed to take place through the cutting process. The change in WRS profile is expected to be influenced from the applied technique as well. Previous investigation carried out by Dattoma et al. compared numerical and experimental results regarding the influence of milling and cutting machining on the WRS of butt welds with different thicknesses [26]. Although the numerical and the experimental results exhibited some deviation from each other, a significant WRS relaxation from both milling and cutting was observed. In some cases, the tensile WRS of the whole plate were even substituted by compressive WRS in the cut specimens as it can be seen in the diagrams of Figure 18.

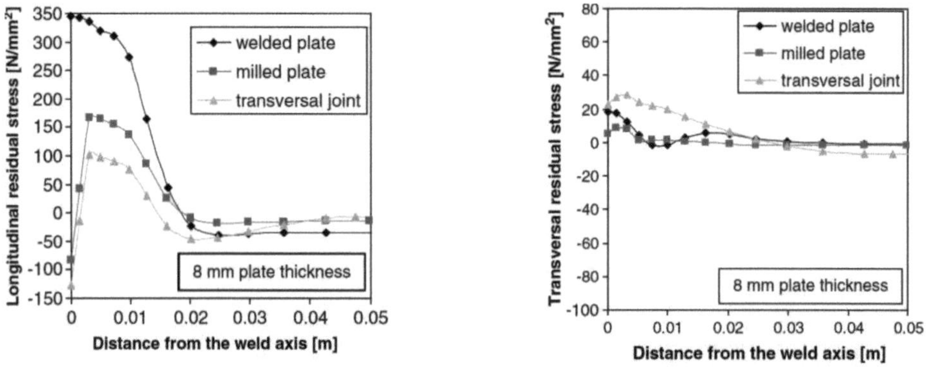

Figure 16: WRS profiles of welded plates with thickness of 8 mm, at as welded, after milling and after cutting states [26]

2.1.2 FE Simulation of HFMI

Increase of fatigue life through the HFMI treatment is affiliated with the introduction of compressive RS (CRS), the increase of local surface hardness and the change of the local geometry in the weld-toe, which leads to reduction of the notch effect [167]. These effects depend on the proportion of energy that is transferred from the HFMI pin to the treated component, through their contact surface, and is transformed to plastic deformation. Plastic strains are introduced, which results in the creation of residual stresses around and beneath the treatment area. Moreover, non-negligible inertia effects may arise due to the rapid introduction of plastic strains. Hence, a precise simulation of the HFMI treatment in the aspect of RS calculation and a safe estimation of the respective fatigue enhancement based on the simulation results, perquisites selection of appropriate analysis type and accurate modelling of the pin energy, the conditions in the contact surface and the material behaviour.

2.1.2.1 Analysis Type for Impact Simulation

During HFMI treatment, the deformation rate of the treated metal is significantly higher than in static events. Therewith, dynamic effects may arise. Different strain rates require different types of analysis, as different aspects of material behaviour have to be taken into consideration respectively. Moradi et al. categorized mechanical testing based on the testing strain rate and reported on the non-negligible aspects in each case as it is shown in Table 2 [125]. These conclusions can be safely extrapolated for numerical investigations as well, as the same physical phenomena have to be simulated. The HFMI treatment is reported to produce strain rates between 200 s^{-1} and 400 s^{-1} [44]. The investigated treatment lies in between the rapid and the impact cases, as they were presented by Moradi et al.

Table 2: Dynamic aspects of mechanical testing, table content found in [125]

Loading Regime	Creep	Static	Rapid	Impact	Ballistic
Typical Time Characteristic [s]	1k	1	20m	10m	10m
Typical Strain rate [1/s]	1m	1	50	1k	1M
Method of Engineering	Creep Rate	Stress Strain	Vibration	Elastic and Plastic	Shock Wave
Inertia Forces	Ignored	Ignored	Considered	Considered	Considered
Thermal	Isothermal	Isothermal	Adiabatic	Adiabatic	Adiabatic
General Stress Levels	Low	Moderate	Moderate	High	High

Therefore, according to initial expectations a full transient structural analysis should be necessary. The dynamic analysis of a finite element system, taking into consideration non-linear material behaviour is governed by the following equation of motion:

$$\mathbf{m}\ddot{\mathbf{u}} + \mathbf{c}\dot{\mathbf{u}} + \mathbf{f}_s(\mathbf{u}) = \mathbf{p}(t). \tag{23}$$

Full transient analysis has been already applied in several cases of HFMI numerical investigations (see [10], [44], [45], [68], [187]). Nevertheless, Boyle et al. have investigated numerically shot peening in [16], the simulation of which exhibits many similarities with HFMI in matters of contact conditions and material behaviour. This previous study has proven that for impact velocity up to 200 m/s quasi-static analysis (Eq. (17)) neglecting inertia phenomena provides results with satisfying preciseness. In the case though, where increased impact velocity of 300 m/s was applied, significant errors arose in the quasi-static analysis, due to neglection of strain rate dependency and interaction of elastic stress waves. Quasi-static analyses of HFMI have been as well carried out in the past (see [103], [115], [142]). However, the results are questionable due to either lacking documentation of the applied methods or to neglection of significant aspects.

2.1.2.2 Material Behaviour

Rate-Dependent Nonlinear Material Behaviour

Yield strength is predominant during HFMI simulation, like in all cases of RS calculations. Le Quilliec et al. have validated with a simplified quasi-static analysis of a notched specimen, whereby friction and WRS were neglected, the significant influence of yield strength on the resulting RS profile [103]. As it is presented in Figure 17 the peak of the introduced compressive stresses was, as initially expected for the modelled bilinear elastic-plastic material behaviour, approximately equal to the nominal yield strength in each investigated case.

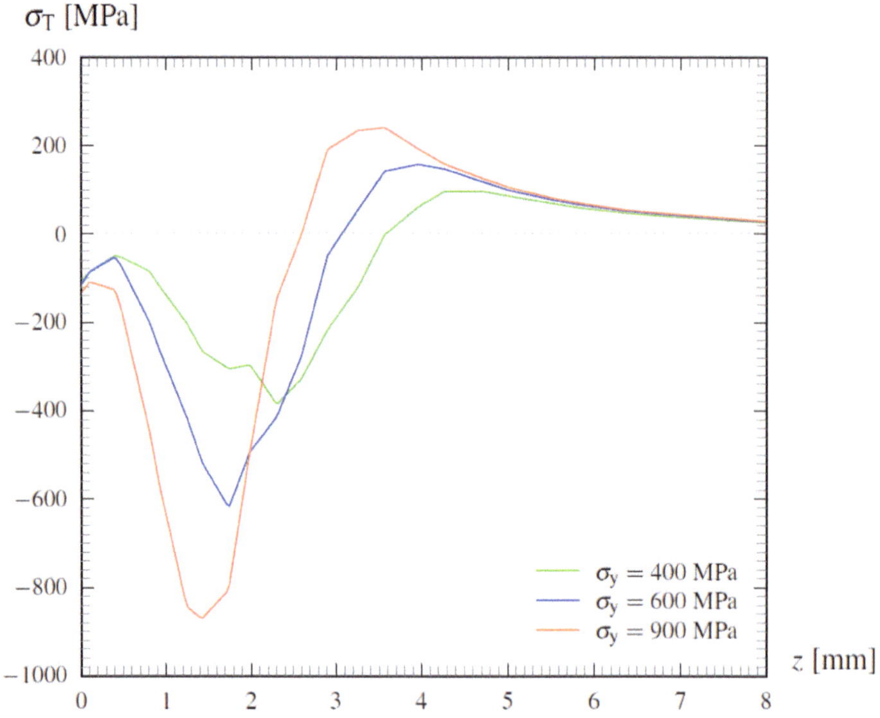

Figure 17: Simulated RS profiles in depth direction for different yield strength values of the investigated material, found in [103] – Component with thickness of 12 mm

Still, the viscoplastic nature of yielding is becoming significant under the high strain rates of HFMI [44]. Jones gathered and correlated data from previous studies on the strain rate dependency of steel at different strain rates and presented the correlation curve of Figure 18 [86]. A rough estimation of the dynamic yield strength under the above-mentioned expected strain rates of HFMI based on that correlation exhibits an increase of more than 100 % of the yield strength. This differentiation of yield strength due to higher strain rates does not allow a direct correlation of the introduced RS with the nominal yield strength, complicating a straightforward calculation of fatigue enhancement in practical applications. According to initial expectations, neglecting the strain rate dependency of the material during an FE simulation of HFMI could lead to erroneous results regarding the calculated RS profile and in extension to a non-conservative estimation of fatigue life. Although it was neglected in several previous studies (see [10], [103], [115], [142] etc.), the necessity of material models considering strain rate dependency of yielding becomes obvious.

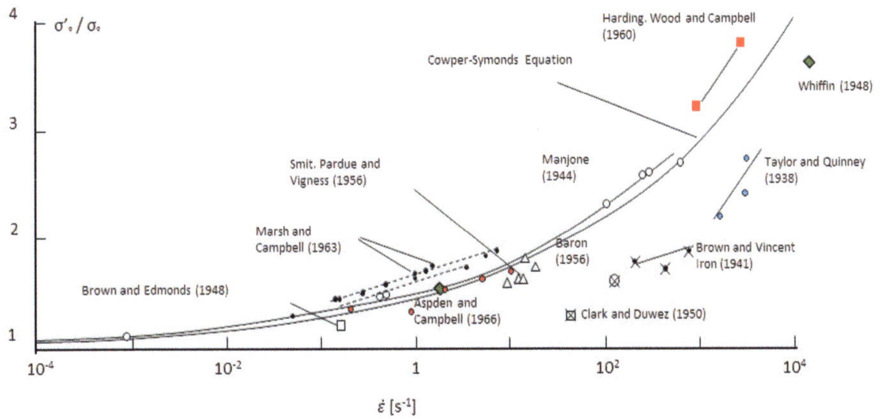

Figure 18: Correlation of static and dynamic yield stress based in experimental data from various studies carried out by Symonds [157], found in [86][12]

The material model described by Equation (24),

$$\sigma_F{}'(\varepsilon_p{}', \dot{\varepsilon}_p, T) = [A + B \cdot \varepsilon_p{}'^{\zeta}] \cdot \left[1 + \Gamma \cdot \ln\left(\frac{\dot{\varepsilon}_p}{\dot{\varepsilon}_{p0}}\right)\right] \cdot \left[1 - \left(\frac{T - T_0}{T_m - T_0}\right)^{\theta}\right], \quad (24)$$

[12] the diagram was redrawn by the author of the present dissertation in order to improve the figure quality

which was presented by Johnson and Cook in 1982 [84], is widely applied in various high strain rate simulations ([46], [78], [101] etc.). Forni et al. investigated the strain rate dependency of structural steel S355 in a series of studies ([47], [48], [49]) and calibrated the Johnson-Cook material model for the investigated alloy based on their experimental results. The parameters calculated by Forni et al. are presented in Table 3.

Table 3: Parameters of the Johnson-Cook model calibrated for S355, found in [49]

A [MPa]	*B* [MPa]	*n* [-]	*C* [-]	*m* [-]
448	782	0.562	0.0247	-

In a more recent study by Cadoni et al. [20] it was further proven that for steel S355 a significant deviation of strain rate dependence is met in tension and compression. Using the Split-Hopkinson set up, tension and compression tests at various strain rates were carried out from 5 s^{-1} up to 8 x 10^3 s^{-1}. The results were compared with static tests for both loading scenarios. A significant deviation in the strain dependent behaviour of steel was documented for the two cases. A summary of the results is illustrated at Figure 19.

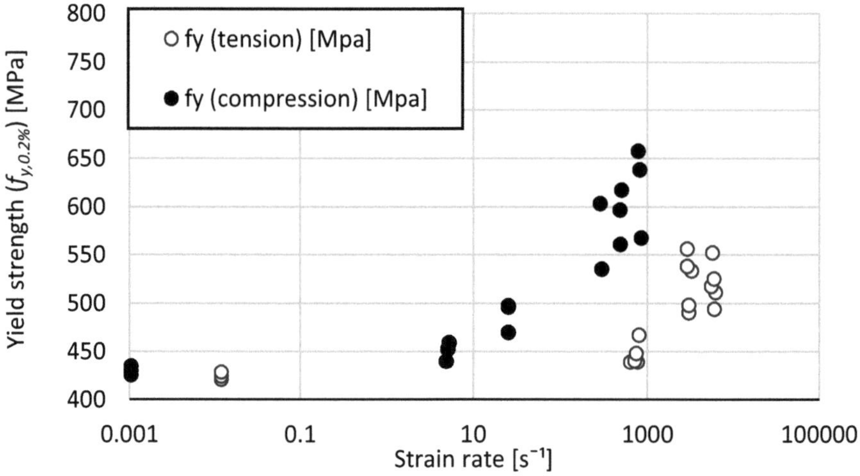

Figure 19: Comparison between dynamic yield strength in tension and in compression, based on a diagram found in [20][13]

[13] the diagram was redrawn by the author of the present dissertation in order to improve the figure quality

Strain Rate Sensitivity

The tests, which were carried out by Johnson and Cook in [84], did not display clearly if there is an interaction between strain rate dependency and temperature, or if they both influence the constitutive relation, independently from each other. The strain rate dependency of each investigated alloy differed from the others', but was significant in any case. In more recent work and due to the evolution of testing techniques, the influence of temperature on strain rate sensitivity was further validated. Strain rate sensitivity (SRS) of metals m_{SRS} is defined by the following Equation,

$$m_{SRS} = \frac{\ln\left({}^{\dot{\sigma}_Y{}'}/{}_{\dot{\sigma}_Y}\right)}{\ln\left({}^{\dot{\varepsilon}'}/{}_{\dot{\varepsilon}}\right)}. \tag{25}$$

It is an index of the strain rate dependency at each temperature level and for each strain rate. Gkatzogiannis et al. [59] compared results from previous studies by May et al. [122], Magee and Ladani [117] and Gupta et al. [65]. SRS of three different alloys, an Al-Mg alloy, a pure 99.5 % Al alloy (denoted as Al 99.5) and AISI 403 stainless steel, were considered. It was concluded that SRS differs significantly for each investigated alloy and an obvious dependence of SRS on temperature was clear in all cases.

Microstructure-Dependence of Strain Rate Sensitivity

Significant discrepancies were observed even in the case of the two batches of the same AL 99.5 with different grain size, the ultra fine grained (UFG) and the conventional-grained (CG) (Figure 20). Even at room temperature, SRS of the two alloys deviates significantly. This time a further significant dependence of SRS on microstructure is confirmed. It is concluded that adoption of rate-dependent material behaviour based on data for another metallic microstructure should be carried out with caution, when the strain rate of an alloy is investigated.

Consequently, further considerations regarding the strain rate dependency of the treated material are required in the case of HFMI. The HFMI treatment is carried out along the weld-toe, practically the HAZ of welded component, where the material behaviour can deviate significantly due to transformed microstructure as it was described above. Extrapolation of the parent material behaviour to the HAZ regarding strain rate plasticity could lead to erroneous results and a consideration of the altered microstructure is required.

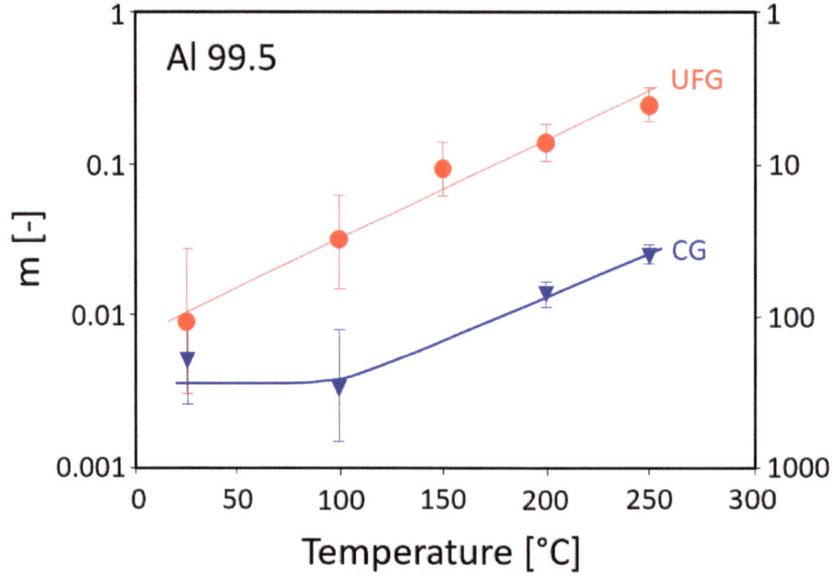

Figure 20: SRS of UFG and CG Al 99.5, *m* corresponds to m_{SRS} of Equation (25), $n = 1/m$ [117][14]

Hardening Behaviour

Material behaviour in cases of reversed loading is expected to influence the HFMI simulation results, as in the case of WRS. During each impact of the HFMI pin on the treated material compressive and tensile stress are expected to arise beneath and around the contact surface respectively. Nonetheless, as the pin moves along the treatment line, consecutive impacts are overlapping. Therefore, areas previously deformed under tension are deformed under compression in subsequent impacts and vice versa.

The hysteretic σ-ε behaviour of metals, which exhibit the Bauschinger effect, is described better by kinematic hardening models. Nevertheless, according to widely accepted engineering knowledge, it provides precise results for small strains only [189]. Mixed hardening has been applied instead in previous numerical investigations of HFMI. Schubnell[15] et al. have investigated the effectiveness

[14] the diagram was redrawn by the author of the present dissertation in order to improve the figure quality

[15] formerly known as Foehrenbach

of kinematic and Chaboche-based [22] mixed hardening models during the simulation of HFMI treatment of a plate from parent material in [44]. Their numerical results were validated based on RS-measurement. The applied Chaboche model provided better results near the treatment surface. Hardenacke et al. have as well previously discussed the influence of hardening behaviour [68]. Isotropic, combined and rate-dependent hardening models produced qualitatively similar results in depth direction, but as in the case of the WRS, significant deviation to the magnitude of the peak residual stresses was met. Zheng et al. [189], Le Quillec et al. [103] and Baptista [10] applied combined, isotropic and kinematic and Chaboche hardening models respectively, while Yuan et al. used usual linear kinematic hardening [187].

Damping

Apart from the deformation behaviour and constitutive law of the investigated material, its damping behaviour should be taken into consideration as well. Precise calculation of RS during nonlinear impact simulations, as in the case of HFMI, perquisites sufficient modelling of the damping behaviour of the investigated material as well, as a portion of the kinetic energy of the pin that is transferred into the component is consumed by the damping mechanisms.

Damping is somewhat an abstract concept. Microscopic mechanisms exist in structures, such as friction between metallic components or opening and closing of microcracks etc., which are responsible for the lack of perfect oscillations in nature. Albeit these mechanisms constitute a property of the structure, which depends on geometry and material, and their influence is evident even for large structures, a direct calculation of their magnitude based on physical characteristics of the structure is not possible. Sophisticated measurements for direct estimation of damping behaviour of existing components/structures are possible. Nevertheless, results from measurements are transferable to analysis of similar structures. Moreover, differentiation of damping behaviour for different levels of external excitation does not allow extrapolation of the damping behaviour from measurements of low excitation to analysis of larger excitation. On the other hand, measurements of energy dissipation, in the case of large excitation, which leads to non-linear deformation, include the plastic deformation energy and complicate even more the evaluation of damping. Hence, predictive modelling of damping, when appropriate measurements are not available, requires an appropriate idealization, which produces reasonable results.

It has been observed that the damping behaviour of a structural system or a component is sufficiently simulated, when it is correlated both with its mass and stiffness [24]. In this sense, Rayleigh's damping model (described thoroughly in [24]), which is a widely applied in numerical analysis and accounts for both mass and stiffness influence was proposed. It is defined as follows:

$$c = a_R \cdot m + \beta_R \cdot k. \tag{26}$$

It was applied during a previous numerical investigation of shot peening, carried out by Meguid et al. [124]. After a *trial and error* procedure, α_R and β_R were set equal to 0.5 and 2 x 10^{-9} respectively. The following viscous damping model,

$$c = 2 \cdot f_0 \cdot \xi \cdot m = D_s \cdot m, \tag{27}$$

was applied instead by Kim et al. for 2D and 3D simulation of shot peening in more recent studies [89], [90]. Different values for ξ between 0 and 0.5 were tested. The resulted RS profiles were almost identical in all cases, arising questions about the negligibility of damping in the investigated type of simulations. A value of 0.5 was proposed for ξ due to numerical efficiency reasons, as it led to the lowest computational time [89]. Such a consideration of damping, based on the model of Equation (27) and applying the same damping factor ξ = 0.5, was made as well by Yuan et al. during numerical analysis of HFMI in [187]. This has been the sole reference to damping in previous modelling efforts of HFMI, known to the author of the present study.

An even more straightforward approach is the application of a global damping factor, i.e. an arbitrary definition of damping matrix C[16] (Eq. (23)), which is applied to the response of all nodes or elements of the investigated component during a FE simulation [113], [114]. Such an approach is usually applied during dynamic simulations, where damping is not predominant for the results, but complete neglection leads to numerical problems.

[16] by giving an arbitrary value to the product $D_s = 2 \cdot f_0 \cdot \xi$

2.1.2.3 Modelling of the HFMI Pin

Material Behaviour of the HFMI Pin

HFMI pin is manufactured by hardened high strength metallic material[17]. Therefore, linear elastic deformation of the pin is expected. Nonetheless, in comparison to the significant plastic strains that are introduced to the treated component, they possibly could be considered negligible without undermining the results. In previous studies of HFMI the pin was simulated as elastic [10], [103], [187], or rigid [44], [68], [142], [189]. Previous investigation of shot peening simulation by Kim et al. [90] has shown that the material behaviour of the pin can significantly influence the calculated RS profiles. Results calculated with rigid, elastic, and elastic plastic behaviour of the spherical shots were compared. The case, wherein plasticity was considered, presented the best agreement with RS measurements, as it is presented in Figure 21. Yet, during shot peening significant deformation of the shots is observed in reality as well, which is not the case for HFMI. Therefore, further investigation is required on this subject.

Modelling the Motion of the HFMI Pin

Two approaches can be adopted for modelling the vertical motion of the HFMI Pin, a displacement-based and a velocity- or acceleration-based. In the former one, the pin is constrained to execute a movement deforming the treated surface up to the defined depth. This option is more straightforward as measuring the depth of the real treatment groove is sufficient for the calibration of the simulated pin movement, which can be carried out with usual laboratory measuring equipment. In previous study of the author [168], the treated area was measured with the optical measurement system LIMESS (Figure 22a) on several butt welds and fillet welds. Three arbitrary lines were selected transverse to the treatment surface of each measured component, as it is illustrated in Figure 22b, and a mean depth of 0.215 mm was documented. The measurement was basis for a displacement based FE simulation of HFMI. A displacement-based approach was adopted

[17] exact information about the pin material is not provided by the manufacturers of the HFMI device due to confidentiality reasons

as well by Le Quillec et al. [103], Zheng et al. [189] and Mangering et al. [115], but strain rate dependency was neglected in all of these studies.

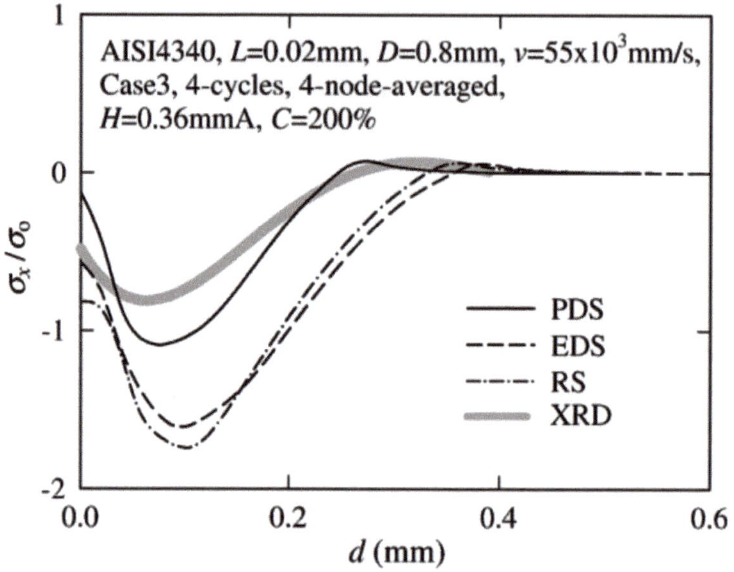

Figure 21: Shot-peening simulation with elastic (EDS), rigid (RS) and plastic (PDS) shots compared with measured RS, found elsewhere [90]

Nonetheless, the selection of displacement-based motion of the pin deviates significantly from physical reality. Unless special measures are taken, a movement under steady velocity during impact and in extension a steady deformation rate is modelled. In reality, the pin is accelerated towards the specimen through an external force and its velocity is reduced as it penetrates the treated surface. Therefore, no steady strain rate is observed. This method is significantly modeller-dependent, as an arbitrary steady velocity for the pin has to be selected, which should produce a strain rate close to the real one.

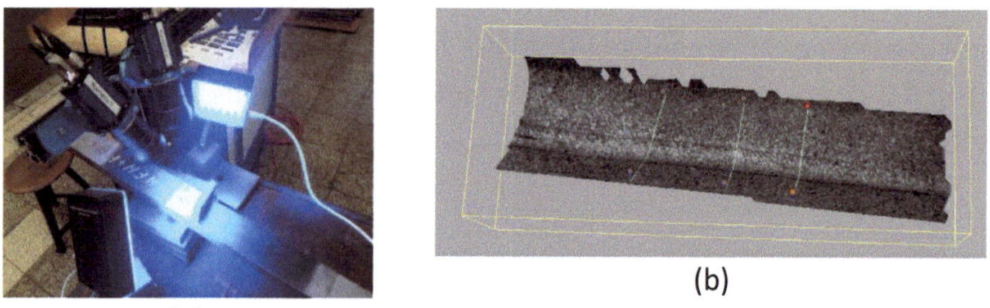

(a) (b)

Figure 22: Measurement of the HFMI groove with the LIMESS measurement system; a) Measurement of a fillet weld; b) 3D depiction of the measured surface and the measurement lines

On the other hand, modelling a velocity-based movement of the pin lies closer to physical reality. The pin is accelerated from a force for a specific time and then the interaction of the system pin – treatment surface during impact can be adequately modelled. Of course, in this case measurement of the input parameters, i.e. the force and the time of its application, is more complicated. Strain gauges were mounted on the pin and the force profile of several impacts was measured by Schubnell et al. [44], a technique that was previously applied by Simunek et al. in [150] as well. Therewith, the contact force and time could be measured. The force was measured for different levels of working air pressure, one of the adjustable settings of the *PITec* device. It was then correlated by means of FE analysis to the impact velocity, which was calculated between 2.1 m/s to 4.2 m/s. A maximum treatment frequency of 120 Hz and a mean travelling speed of 7 mm/s (42 cm/min) were as well documented in [44] and [146]. Therewith, a mean traveling step of 0.06 mm between consecutive impacts can be assumed. It was further observed that the strain gauges' indication between consecutive impacts was non-zero. This is attributed either to secondary impact or to reflection of stress waves inside the pin. In any case, these secondary force indications were neglected in subsequent analysis and the pin was accelerated with a specific force until the impact force was achieved. The contact force was then used as validation between measurements and the FE Model.

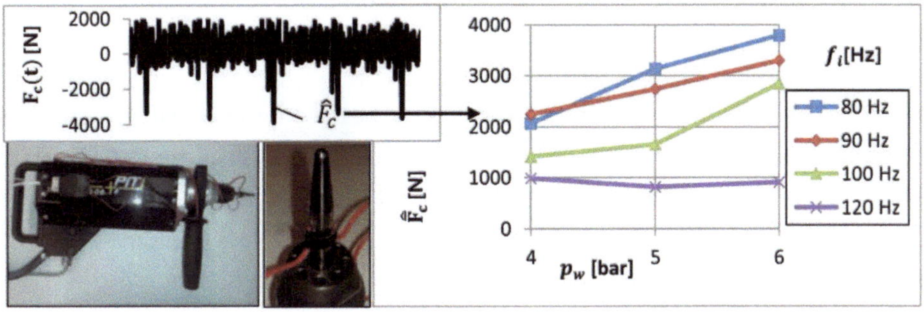

Figure 23: Measuring the contact force of the *PITec* device and its correlation to working frequency, found in [44]

Similar measurements were carried out by Schubnel and Gkatzogiannis on a *HiFIT* device as well in the framework of *HFH-Simulation* [146]. A contact time between the accelerating device and the pin of 0.1 ms was calculated. For an impact frequency range of 209 Hz to 282 Hz, a contact force of 4100 N and 5500 N was measured. Once again, through a correlation by means of FE analysis the contact force was correlated to an impact speed of 3 m/s to 4.5 m/s respectively. A higher mean travelling speed of 48 mm/s (288 cm/min) was documented for *HiFIT* in comparison to the 7 mm/s (42 cm/min) of *PITec*. The mean travelling step between consecutive impacts is nonetheless significantly higher and equal to 0.22 mm due to the increased frequency in this case.

A velocity-based approach was adopted by Yan et al. [187], while the two methods were compared by Schubnell et al. [44] and Hardenacke et al. [68]. Actually, it was concluded by the former one that the displacement-based approach produced results showing better agreement with measured RS. The authors recommend though that the displacement-based model should be applied with strain rate dependent material behaviour, but the applied displacement-rate of the pin, which would affect the rate dependent material behaviour, is not reported.

Finally, regarding the moving velocity of the pin, a step of 0.4 mm and 0.3 mm was proposed in [44], [68], and [150] and [187] respectively. Both values are larger than the real overlap between consecutive impacts but it was mentioned in the above studies that this deviation had a negligible effect on the results. In a more recent study [145] though, the influence of coverage has been investigated thoroughly. It was suggested based on RS measurements that an increased overlap up to a threshold value leads to a higher magnitude of transverse RS. Above this threshold point though, no significant change on the RS field was observed.

Modelling of Contact Conditions

Appropriate definition of contact conditions between the HFMI Pin and the treated surface is required in order to ensure a close to physical-reality simulation of contact problems. Contact dissipates part of the pin's kinetic energy, prior to its transfer in the component in the form of stress waves. Coulomb's friction model (Eq. (28), see [114]), which is described by Equation (48)

$$F_f = \mu \cdot F_N, \tag{28}$$

is widely applied for metallic materials and has been applied for the simulation of HFMI and shot peening in the past. The coefficient μ takes various values depending on the nature of the surfaces, which slide under friction. It is usually characterized as static or dynamic, when it describes the state at the beginning or during sliding respectively. Generic values for different surface qualities and for both cases, the static and the dynamic, are proposed in [6]. A selection from the proposed values is presented in Table 4.

Table 4: Typical values of Coulomb friction for dry contact, found in [6]

Materials	Static	Dynamic
Hard steel on hard steel	0.78	0.42
Mild steel on mild steel	0.74	0.57
Aluminum on mild steel	0.61	0.47
Aluminum on aluminum	1.05	0.40

2.1.2.4 Modelling of Boundary Conditions

As it was described above for the case of FE simulation of welding, modelling of restraints applied to a real component can have a significant influence on calculated RS. The influence of HFMI treatment in a component is very localized, being only some mm in all directions as it was shown by previous RS measurements [44]. One could assume that restraints, which are applied tens or hundreds of millimetres away from the treatment area, would have a negligible effect on the calculated RS, depending on the bending stiffness of the treated component as well. This assumption could stand, when a simulation of a larger component takes place.

However, precise FE simulation of HFMI perquisites a very fine mesh, lower than 0.1 mm as it was shown in previous analysis [168], a condition that rises significantly the computational time. Hence, a full-scale analysis of a large component would not be possible in rationally short time. Only smaller components or parts of larger components can be modelled. In these cases, a significant effect of the boundary conditions could be observed. For small components, realistic modelling of clamping-down mechanisms or of restraint-free condition as in the case of welding is necessary. For modelling parts of a larger structure, appropriate boundary conditions should be selected. Mechanical restraints from the wider areas of the larger structure to the small, simulated part should be modelled realistically. Acoustic impedance of the boundaries should be nullified, in order to approach the elastic stress wave propagation of the larger component. No special reference to the selection of appropriate boundary conditions is made in previous analyses of HFMI. In most cases, nodal displacements at the bottom of the investigated component were fixed to all directions without further explanation.

2.2 Analytical Investigations for the Calculation of the Dynamic Yield Strength during a Spherical Impact

Hertzian theory of elastic contact has been the basis for analytical solutions of impacts since its evolution more than 130 years ago [73]. In 1998, Lim and Stronge proposed such an analytical solution for the calculation of a rigid cylinder's impact on an elastoplastic-half space [110]. In their study, this impact was analysed in three consecutive phases, the elastic, whereby Hertzian theory is applied, the elastic-plastic, whereby elastic and plastic behaviour coexist in the area beneath the contact and finally, the fully plastic phase. Their solution beyond yielding was based on the *cavity* model, which was initially proposed by Hill [75] and later simplified by Johnson [83]. This model assumes the symmetrical expansion of a spherical cavity in an elastic-plastic material so that compatibility between the volume of displaced material and the accommodated radial elastic expansion is ensured during the contact.

Classical Hertzian theory [73] refers to elastic bodies. In the present case, the impacting sphere is regarded rigid ($E \rightarrow \infty$). Therefore, the physical quantity defined by Hertz as E^* and given by Equation (29),

$$\frac{1}{E^*} = \frac{1 - {\nu_{sphere}}^2}{E_{sphere}} + \frac{1 - {\nu_{plate}}^2}{E_{plate}}, \tag{29}$$

will hereinafter be given by the following equation

$$\frac{1}{E^*} = \frac{1 - {\nu_{plate}}^2}{E_{plate}}. \tag{30}$$

According to Hertzian theory ([73], [110]) the indentation depth of a rigid sphere in an elastic half space and in the elastic regime is calculated as follows:

$$\delta = \left[\frac{9 \cdot F^2 \cdot (1-\nu^2)}{16 \cdot R \cdot E}\right]^{1/3}. \tag{31}$$

Under the assumptions of infinitesimal strains in the elastic regime and continuous contact between the indenter and the half-space, the contact area between the sphere and the plate is circular and the contact force is correlated to the mean pressure according to the following Equation (32),

$$F = \pi \cdot a^2 \cdot p_m. \tag{32}$$

The contact half width α in the elastic regime and therefore during yielding as well, is given by Equation (33):

$$a^3 = \frac{3 \cdot F \cdot R \cdot (1-\nu^2)}{4 \cdot E}. \tag{33}$$

When Lim and Stronge investigated the elastic-plastic impact of the rigid cylinder on a half-space [110], they noticed that after the end of the indentation and unloading and as long as full plastification has taken place, the crater's half width α_c after full plastification is correlated with the contact half width at yielding according to Equation (34),

$$a_c^2\left(\frac{1}{R} - \frac{1}{R'}\right) = \frac{a_y^2}{R}, \tag{34}$$

where *R* and *R′* is the radius of curvature of the contact area prior and after unloading. Prior to unloading *R* is equal to the radius of the indenter, as full contact is assumed at all times. This latter equation was extracted assuming that the contact width at maximum compression α_c is equal to the width of the residual indentation i.e. the elastic spring back is negligible, the residual indentation has a final curvature *R′*, which is slightly larger than *R* due to elastic recovery and finally, no piling up or sinking in takes place during the indentation of the plate.

Johnson, analysing the indentation of an elastic - perfect plastic and incompressible material by a rigid sphere, proposed in [83] that the average pressures $p_m{}^Y$ and $p_m{}^P$ at initiation of yielding and at full plastification are respectively correlated with the material yield stress according to the following equations:

$$p_m{}^Y = 1.1 \cdot \sigma_Y, \tag{35}$$

$$p_m{}^P = 3.072 \cdot \sigma_Y. \tag{36}$$

These empirical equations were subsequently validated for steel by numerical and experimental investigations (see for instance [159]). Yet, the diagram of Figure 24, found in [82], reveals that these relationships between average pressure and yield strength, which mark the transition from the elastic to the elastic plastic and from the latter one to the perfect plastic regime, are dependent on the yield strength itself. The provided curves are proposed for mild steel. Nevertheless, the bulk of the experimental results of the diagram, illustrated with circular marks, were deployed by a previous work of Tabor [159], whereby a steel of yield strength of 770 MPa was investigated. In a more recent study [81], Jackson and Green validated this dependency of the average pressure to yield strength on the yield strength itself. Five different materials with yield stress of 210 MPa, 560 MPa, 911 MPa, 1265 MPa and 1619 MPa were investigated and the factor of Equation (36) was found to be 2.8, 2.6, 2.5, 2.4 and 2.3 respectively. Similar values are validated from the diagram of Figure 24, if a yield stress of 355 MPa is assumed for the mild steel and the logarithmic axis of non-dimensional strain is adjusted for the increased yield stress based upon this assumption.

Finally, Johnson [82] defined the coefficient of restitution during an impact:

$$e^2 = \frac{v'^2}{v^2}, \tag{37}$$

where v and v' are the initial and restitution velocities respectively [82]. He also proposed that for an inelastic impact (elastic – perfect plastic behaviour of the target) of a rigid sphere at moderate impact speeds (v around 5-100 m/s) it can be correlated to the dynamic yield limit as follows:

$$e \approx 3.8 \cdot \left(\frac{\sigma_Y{}'}{E^*}\right)^{\frac{1}{2}} \cdot \left(\frac{\frac{1}{2} \cdot m_{sp} \cdot v'^2}{\sigma_Y{}' \cdot R^3}\right)^{-\frac{1}{8}}. \tag{38}$$

A similar relationship had been earlier proposed by Tabor [159] as well. Equation (38) was extracted by Johnson from the following equation

$$e^2 = \frac{3 \cdot \pi^{5/4} \cdot 4^{3/4}}{10} \cdot \left(\frac{p_d}{E^*}\right) \cdot \left(\frac{1/2 \cdot m_{sp} \cdot v'^2}{p_d \cdot R^3}\right)^{-1/4}, \tag{39}$$

based on the assumption that under dynamic events the condition $p_m \approx 3 \cdot \sigma_Y'$ is still valid, when the regime of perfect plasticity is entered.

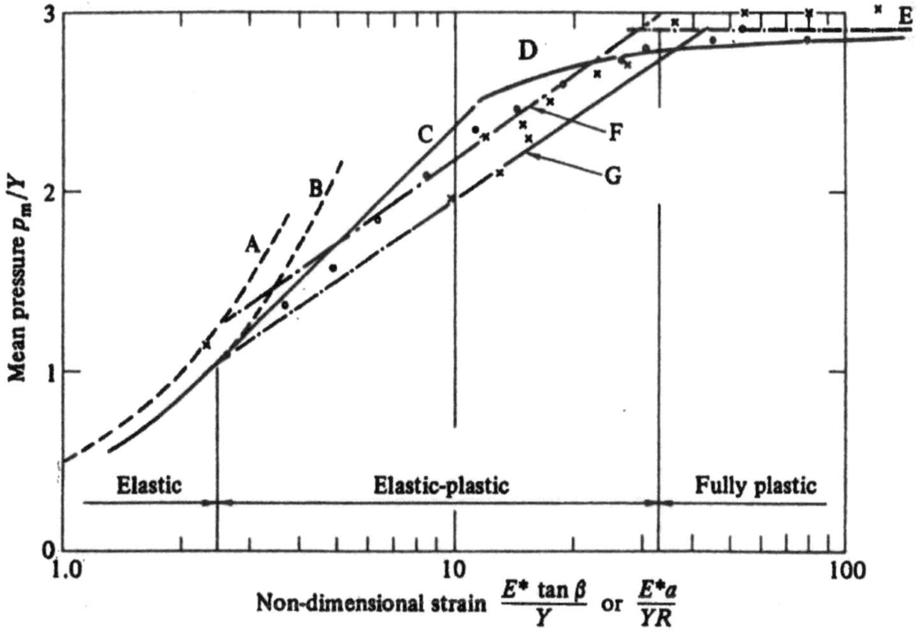

Figure 24: Relationship between mean pressure p_m and yield stress (marked as Y) and non-dimensional strain found in [82]

Implementing the above-presented set of equations, a phenomenological estimation of the dynamic yield stress during an impact of a sphere on a semi-infinite plate, based on measurements of either the residual trace or the rebound velocity, can be conducted. According to Johnson [82], the above-presented relationships are part of the shallow indentation theory, which is valid up to impact velocities of 100 m/s. Nevertheless, according to Lime and Stronge [110], as a discontinuity of contact half-width with increasing indentation is created at the transition area between elastic-plastic and fully plastic behaviour, calculations based on measurement of the crater should be avoided for cases were the maximum indentation δ_c is only slightly larger than that required for entering full plasticity.

2.3 RS Profiles Introduced by the Impact of a Metallic Sphere

Numerous previous studies investigated impact problems such as HFMI, shot peening or other processes that fall into the generic category of a metallic rigid sphere impacting on a deformable metallic plate ([16], [63], [64], [90], [130], [115], [124], [166] etc.). RS profiles for different sphere diameters, boundary conditions, impact velocities and materials have been extracted either numerically or experimentally. Although the uniqueness of each investigated case has been underlined above, in most cases significant qualitative similarities are met. In order to enable a better overview and a more straightforward validation of the results, representative profiles proposed by Boyce et al. [16] for an ideal impact of a metallic sphere, by Guagliano [63] for shot peening and by Nitschke-Pagel et al. [130] for HFMI, are presented respectively in Figure 25, Figure 26 and Figure 27 below.

Compressive stresses are met near and underneath the impact crater of all presented examples. In all these three cases and for the stresses to all directions, longitudinal, transversal or hoop, the peak of the compressive stresses is met underneath the impact crater at a distance up to 1 to 1.5 times the crater radius. According to Foehrenbach et al. though [44], impact velocity should as well influence the depth of the peak. On the surface the peak of the compressive stresses is met at the rim of the crater. It is becoming evident by the diagrams of Figure 26 and Figure 27 that increased intensity or impact speed and application of consecutive impacts introduce a shift of the RS profile without qualitatively changing the distribution. Similar results were presented in the majority of previous studies, which are omitted for the sake of space, either by means of experimental measurements or by numerical analysis. Therefore, initial expectations dictate a similar profile for the current investigations of HFMI, when an unnotched specimen of parent material is investigated.

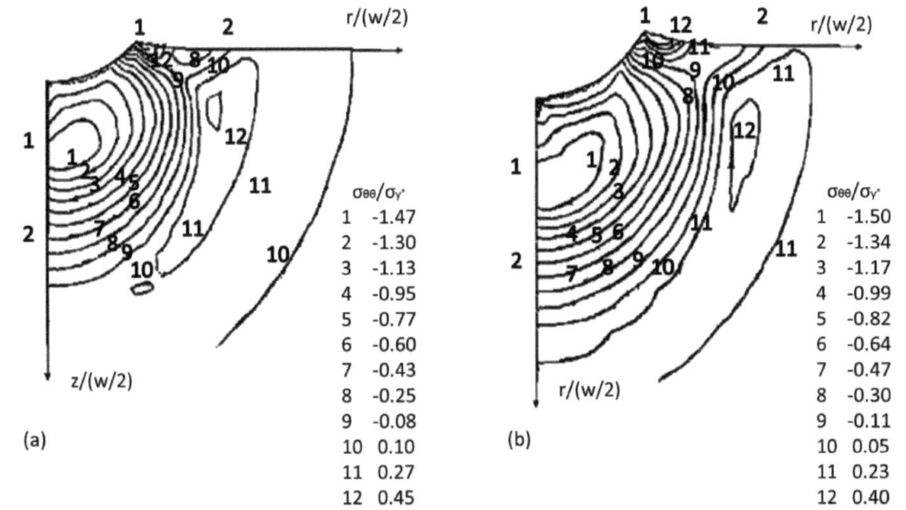

Figure 25: Contour of hoop stresses predicted by a FE model by Boyce et al. [16] for the impact of a rigid sphere with 200 m/s (a) and 300 m/s (b) on a plate of Ti-6Al-4V alloy – Stresses and distance from crater's centre are normalized to the static yield strength and the crater diameter respectively – *w* is the diameter of the crater[18]

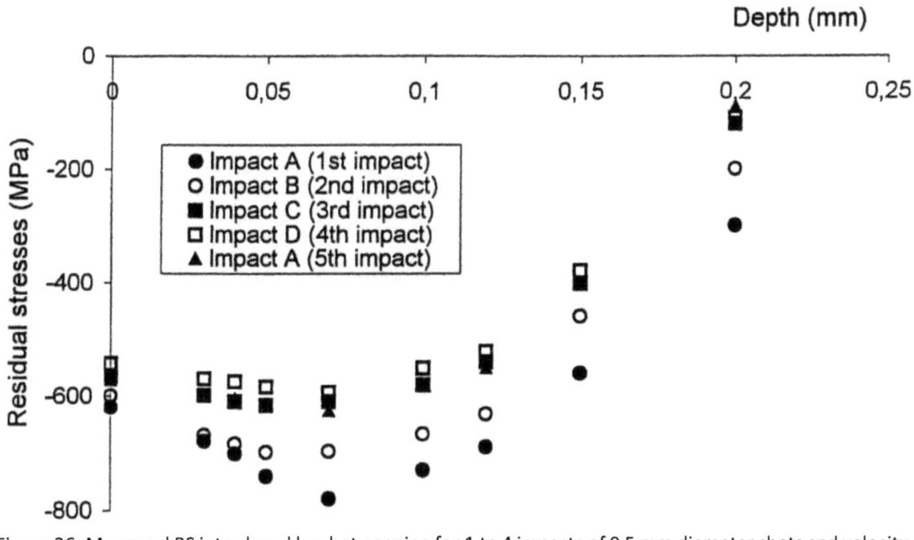

Figure 26: Measured RS introduced by shot peening for 1 to 4 impacts of 0.5 mm diameter shots and velocity of 100 m/s, a crater diameter of 0.1 mm is calculated based on figures found in the literature source, found in [63]

[18] the figure was partially redrawn by the author of the present dissertation in order to improve the quality

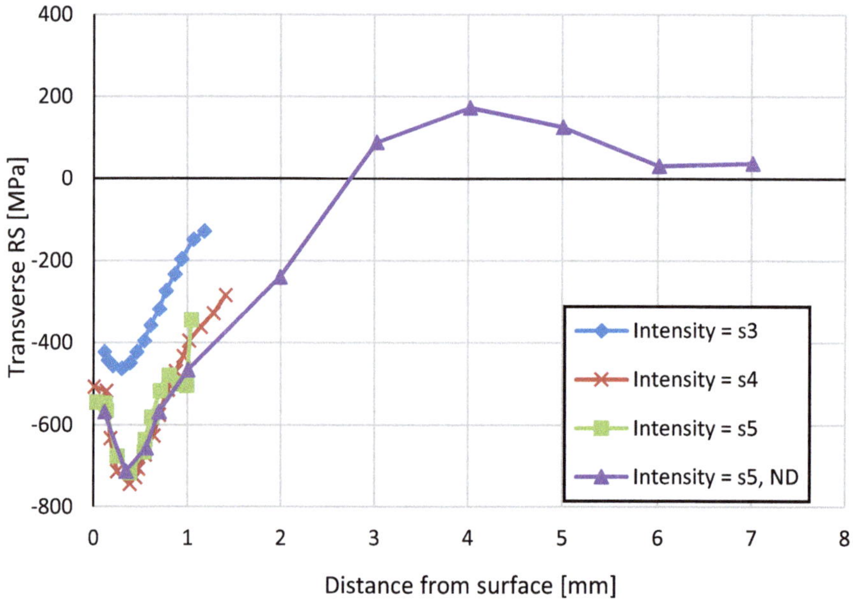

Figure 27: Transverse residual stress through-depth profiles in UIT-treated fields for variating treatment intensity and for a pin diameter of 4.8 mm in the base material S690, crater swallower than 0.5 mm, based on a diagram found in [130]

3 FE Simulation of Welding

3.1 Methodology

Simulations of welding, which were carried out in the framework of the present dissertation, were based on the above-presented theoretical background. Each investigated case acted as a milestone towards the completion of the present task, the FE simulation of welding, fulfilling the conditions that were set at the beginning of the present research study regarding straightforwardness and preciseness.

Single-pass butt welds were simulated as a first step towards the development of the present approach due to their simple geometry and the reduced required modelling, as the influence of a single thermal cycle has to be considered. Validation was initially based on measurements, which were found in literature. The classical example presented by Anderson back in 1978 - component A - along with a newer study from Caron et al. - component B - were used as reference ([3] and [21] respectively). These two cases were selected, due to the full documentation of the welding procedure along with WRS measurements provided by the authors. The components were remodelled with the current approach, applying identical welding parameters and geometry. Some of the results were presented in earlier work of the author in [53], [59] and [92].

The results regarding component A were compared with other efforts of simulating the same component by Anderson himself [3] and Lindgren in [111] as well. Results from the simulation of component B were compared with the respective simulation carried out by Caron et al., using the specialized FE software *SYSWELD* [158]. Valuable conclusions regarding several practical aspects of weld modelling were drawn and applied in subsequent simulations.

Next step was the extension of the present approach for the case of multi-pass welding of austenitic steels [56] and the simulation of aluminium alloys [93]. Nevertheless, these investigations although relative to the filed, are only outlined and a detailed presentation is excluded from this dissertation, as they do not directly contribute to the present research goal.

The presented approach was finally applied, in order to simulate plates with transversal stiffeners (fillet welds) welded on both sides from the research project *HFH-Simulation* [146], wherein RS measurements were carried out in as-welded and HFMI-treated state (see [146]). Goal was the calculation of the WRS field so that it could be applied as input for the subsequent HFMI simulation.

Although modifications based on the available information and improvements in modelling were carried out after each investigation, the backbone of the approach remained unaltered, as it was initially presented in [92]. The physics fields and the predominant interactions that are presented in Figure 4 were taken into consideration. FE commercial software *ANSYS* was applied for all weld simulations.

3.1.1 Thermal Transient Analysis

A transient thermal analysis was initially carried out in order to calculate the thermal history of each node, based on Equation (4). *ANSYS* 8-node solid finite elements *SOLID70*, which allow the simulation of heat conductivity and heat generation, were applied [4]. The element temperature is calculated at each time step based on its shape functions, from the temperatures of its nodes (see Appendix B).

Goldak's heat source was applied, using an algorithm, which was provided by the company CADFEM. It was applied in such a way, that the centre of the source was moving at each solution step along the welding line, for a distance equal to the welding speed multiplied by the time between the consecutive solution steps. Power of the heat source is calculated according to Equation (3). Values for the coefficient of heat source η proposed by Dupont et al. were applied based on the investigated weld type (see [32]), unless otherwise stated below. Heat losses were simulated according to Equation (7), using a common coefficient for heat losses h_T, as described above. Values for the coefficient were found in [140] and they are presented in Table 5. Thermal material properties were assumed independent of the microstructural transformations in all cases. Therefore, thermal properties of the parent material were applied in the thermal transient analysis. Ambient temperature was assumed equal to 20 °C in all cases. In each solution step, information regarding elements whose temperature was exceeding the melting point was documented. After completion of the transient solution the double parameter (T_{max}, t_{85}) of each element in the HAZ and FZ was calculated based on the temperature history of its nodes.

Table 5: Applied values for the total coefficient of heat losses, found in [140]

Temperature [°C]	20	100	450	850	1550	2350	3000	10000
h_T [W/(m^2 · K)]	4.7	22.8	48.2	117.9	396.9	1082.3	2032.9	58214.0

3.1.2 Microstructural Modelling

Microstructural modelling was carried out based on the approach presented in [92] and described in the theoretical background. A bilinear elastic – plastic material behaviour was assumed in most cases, unless otherwise referred. Parent material properties for the butt welds were taken from literature, while for the transversal stiffeners respective measurements were carried out in the framework of the project *HFH-Simulation* [146]. *A1* and *A3'* were calculated using Equations (9) and (10), (11) based on the chemical composition of each investigated alloy. Bilinear elastic – plastic CMMs were built for different levels of T_{max} and for different cooling rates t_{85}. The former influenced the austenitization percentage and the later the final transformed microstructure, which unless otherwise stated, was calculated from appropriate CCT diagrams found in literature. The dilation behaviour and yield strength of the CMMs was calculated by applying linear mixtures law, with Equations (15) and (19) respectively. Yield strength of the individual phases was found in [74]. Unless otherwise mentioned, coefficient of thermal expansion of austenite and the rest of the phases was assumed equal to 16×10^{-6} and 12×10^{-6} respectively and tangent modulus *H* was assumed same for all phases. Whenever necessary, TRIP was taken into consideration according to Karlsson's approach, as it was described earlier.

3.1.3 Static Structural Analysis

A quasi-static structural analysis (see Equation (17)) was carried out based on the thermal history of the nodes, which was calculated during the transient thermal analysis. *ANSYS* 8-node solid finite elements *SOLID185*, which have stiffness and allow simulation of stress and thermal loading, were applied for this analysis. The respective shape functions are provided in Appendix B. Same mesh from the transient thermal analysis was used, changing only the element type and retaining previous geometry.

In each solution step, thermal strains were applied as loading according to Equation (18) based on the temperature history of the thermal simulation. During heating, parent material parameters were assumed along the whole investigated component. When T_{max} of each element was reached, a CMM was assigned to it. Assignment was carried out by selecting through appropriate algorithm the CMM with values of the double parameter closest to the calculated ones of the element (T_{max} and t_{85}), by means of linear interpolation. At each load step, the elements whose temperature exceeded the melting point were deactivated, leading to erasure of previous strain history. The elements were reactivated when their temperature reached the solidus point once again. Therewith, the addition of the filler material and its influence on the WRS could be simulated. Von Mises flow rule was applied (Equation (17)).

Table 6: Applied values for the individual microstructural phases of steel, found in [74]

Temperature [°C]	Static Yield Strength [MPa]			
	Ferrite and Perlite	Bainite	Martensite	Austenite
1500	5	5	5	5
1205	11	5	14	12
877	57	29	78	49
812	79	42	120	62
713	86	73	264	81
420	218	237	783	136
200	285	271	880	200
20	431	277	1008	245

3.2 Single-pass Butt Welds

3.2.1 Investigated Components

Anderson presented in [3] the investigation of the single-pass submerged arc welded[1] component, which is illustrated in Figure 28 (component A). The X-grooved butt weld of Swedish structural steel HT36 with dimensions of 2000 mm x 500 mm x 25 mm was welded with an electric power of 98 kW and a welding speed of 25 mm/s (150 cm/min; 3.92 kJ/mm gross heat input). The geometry of the cross section is presented in Figure 29.

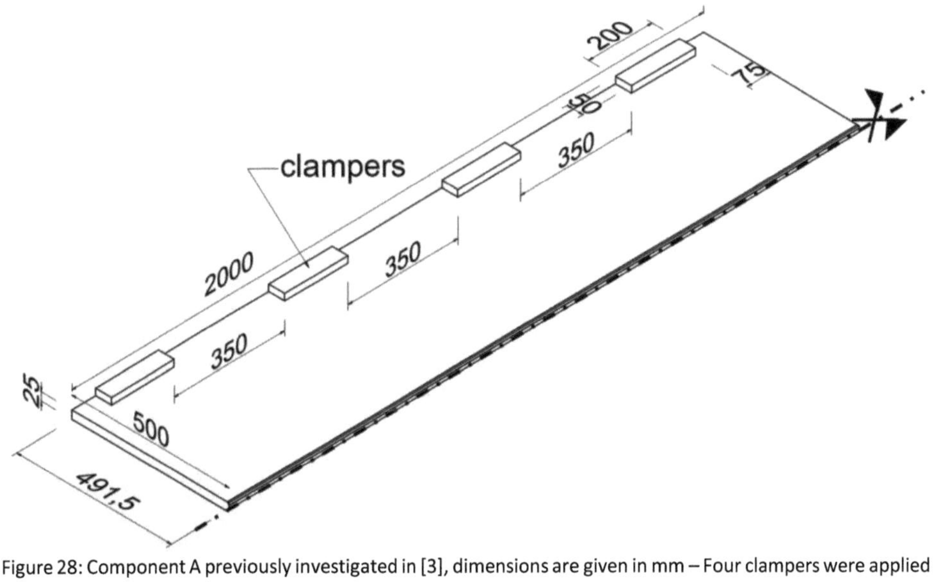

Figure 28: Component A previously investigated in [3], dimensions are given in mm – Four clampers were applied on each side, the applied symmetry condition is presented as well

Caron et al. investigated in [21] a single-pass V-grooved butt weld of steel S355 [29], which is presented in Figure 30 (component B). The component with dimensions of 500 mm x 200 mm x 5 mm was welded with an electric power of 7.934 kW and a welding speed of 6.7 mm/s (40 cm/min; 1.18 kJ/mm gross heat input).

[1] three electrodes welding consecutively, one behind another

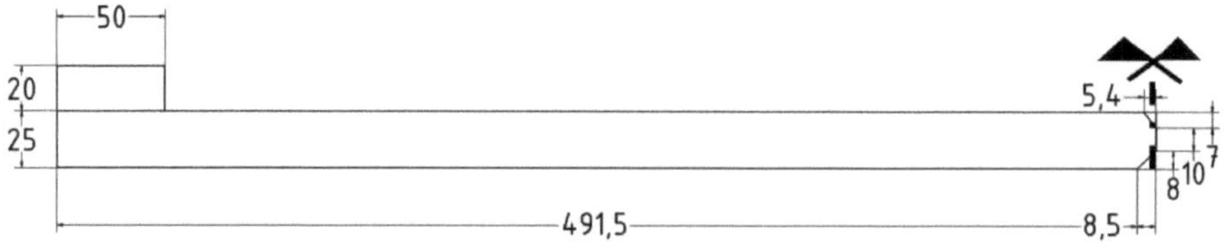

Figure 29: Cross section of component A, dimensions are given in mm – The applied symmetry condition is presented as well[2]

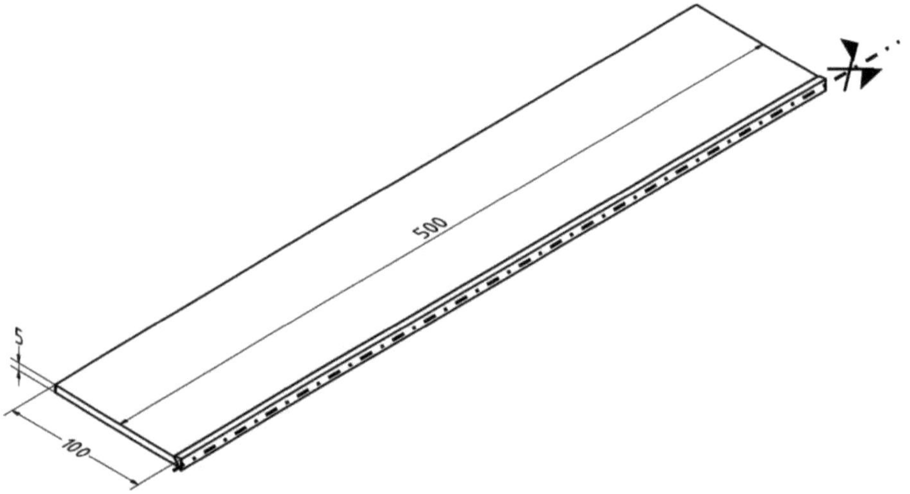

Figure 30: Component B previously investigated in [21], dimensions are given in mm – No restraints during welding, the applied symmetry condition is presented as well

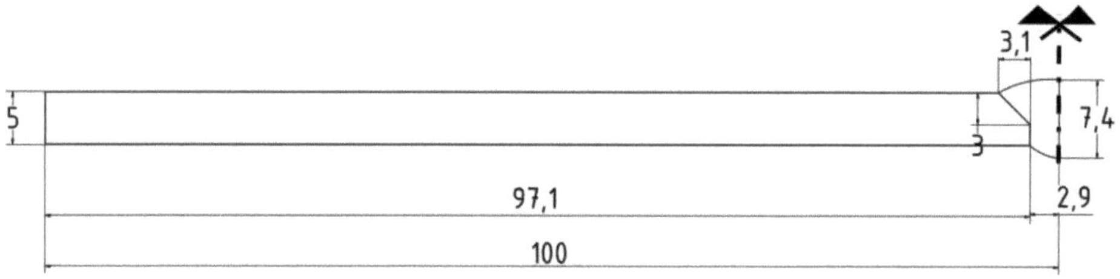

Figure 31: Cross section of component B, dimensions are given in mm – The applied symmetry condition is presented as well

[2] the component was X-grooved as it would be welded with 3 passes of a single electrode – with the applied submerged arc welding technique with 3 electrodes welding consecutively, one behind another, it was possible to weld with a single-pass the middle and upper seam, the final geometry of the weld section area is illustrated better bellow (see Figure 35), more information is provided in [3]

3.2.2 Material Modelling

3.2.2.1 HT36

Applied parent material properties of steel HT36 were provided by Andersson in his original study [3]. They are presented in Table 7. The chemical composition of HT36 and the results from the analytical calculation of the austenitization temperatures *A1*, *A3* and *A3'* are presented in Table 8. CMMs were built based on the CCT diagram of HT36, which is presented in Figure 32, and the material properties of the individual phases (see Table 6). For the simulation of component A, 6 CCMs were built, with each CCM being characterized by one pair of T_{max} and t_{85} (see Table 9). Exemplary, evolution of chemical composition and the respective mechanical behaviour of CMM 2 as a function of temperature are provided in Table 10. For the sake of shortness, the rest of the CMMs are provided in Appendix A.

Table 7: Temperature-dependent material parameters of HT36 found in [3]

ρ [kg / m³]	T [°C]	c [kJ / (kg · K)]	K [W / (m · K)]	σ_y [MPa]	E [GPa]	H [GPa]
7800	0	0.4	40	355	220	2.2
	200	0.5	40	320	200	0
	600	0.6	40	102	131	0
	700	0.9	36	84	98	0
	850	1.3	26	56	49	0
	900	0.6	26	47	32	0
	1500	0.8	33	5	0	0

Table 8: Chemical composition of HT36 and analytically calculated temperatures of austenitization (Equations (9), (10), (11))

Chemical composition HT36				*A1* [°C]	*A3* [°C]	*A3'* [°C]
C	Si	Mn	Nb			
0.13	0.2	1.57	0.024	712	801	1250

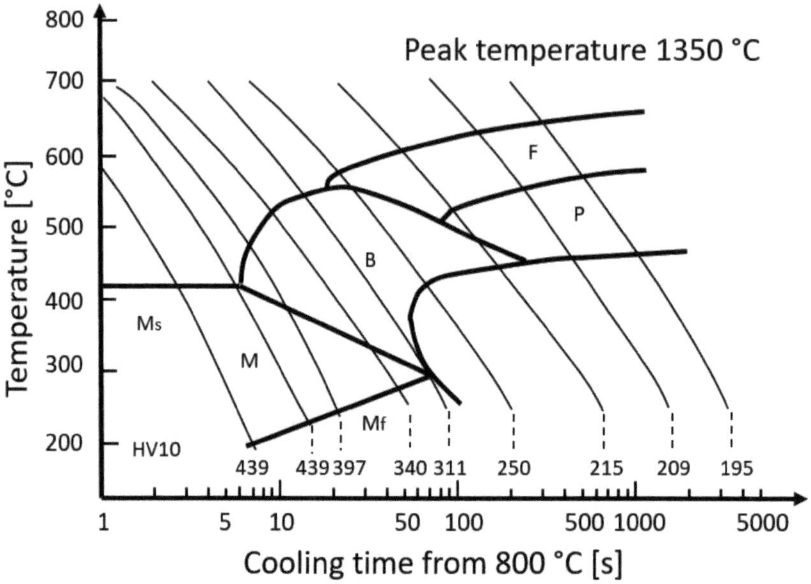

Figure 32: CCT diagram of steel HT36, from [3][3]

Table 9: Double parameter of the cooling material models for the simulation of HT36, based on the *Gkatzogiannis Approach* [92]

CMM	2	3	4	5	6	7
T_{max} [°C]	950	1150	1350	950	1150	1350
t_{85} [s]	10	10	10	35	35	35

[3] the diagram was redrawn by the author of the present dissertation in order to improve the figure quality

Table 10: CMM 2 (T_{max} = 950 °C, t_{85} = 10 s) modelled for the simulation of microstructural evolution of HT36, based on the *Gkatzogiannis Approach* [92]

T	Microstructure					σ_{yi} [MPa]					σ_y	α^{se}
[°C]	PM	AY	FE, PE	BA	MA	PM	AY	FE, PE	BA	MA	[MPa]	[-]
950	63 %	37 %	0 %	0 %	0 %	53	38	38	38	38	43	18
520	63 %	37 %	0 %	0 %	0 %	114	177	363	483	175	152	18
320	63 %	7 %	0 %	30 %	0 %	148	206	445	569	270	313	15
270	63 %	0 %	0 %	30 %	7 %	111	250	485	612	188	307	15
20	63 %	0 %	0 %	30 %	7 %	200	258	490	620	350	411	15

3.2.2.2 S355

Temperature-dependent parent material properties of steel S355 were found in [34]. Values for the tangent modulus provided by Byfield et al. were adopted [19]. The applied material properties are presented in Table 11. The chemical composition of S355 batch, which was investigated by Caron et al. [21], and the results of the analytical calculation of the austenitization temperatures *A1*, *A3* and *A3'* are presented respectively in Table 12 and Table 13. 6 CMMs were modelled based on the CCT diagram of S355, which is presented in Figure 8, and the material properties of the individual phases (see Table 6). The values of the double parameter T_{max} and t_{85} characterizing each CMM are provided in Table 14. Exemplary, evolution of chemical composition and the respective mechanical behaviour of CMM 2 are provided in Table 15. For the sake of shortness, the rest of the CMMs are provided in Appendix A.

Table 11: Temperature-dependent material parameters of S355 found in [3][4]

ρ [kg / m^3]	T [°C]	c [kJ / (kg · K)]	K [W / (m · K)]	σ_y [MPa]	E [GPa]	H [GPa]
7800	0	0.4	54	355	220	2.2
	200	0.5	47	355	220	2.2
	600	0.8	40	167	65	0.3
	700	1.0	31	82	28	0
	850	1.5	27	30	17	0
	1000	0.7	27	14	9	0
	1500	0.7	27	5	9	0

Table 12: Chemical composition of S355 batch used by Caron et al. in [21]

C	Si	Mn	P	S	Cr	Mo
0.14	0.2	0.67	0.00823	0.012	0.033	0.01
Al	Nb	Ti	V	Cu	W	Ni
0.0354	0.005	0.0023	0.00393	0.0221	0.01	0.0354

Table 13: Calculated austenitization temperatures of S355 (Equations (9), (10), (11))

A1 [°C]	*A3* [°C]	*A3'* [°C]
721	843	1035

Table 14: Double parameter of the cooling material models, which were given into the algorithm for the simulation of S355, based on the *Gkatzogiannis Approach* [92]

CMM	2	3	4	5	6	7
T_{max} [°C]	826	931	1276	826	931	1276
t_{85} [s]	32	32	32	85	85	85

[4] the values of σ_y and E at 1500 °C were arbitrarily set equal to non-zero values due to numerical reasons

Table 15: CMM 2 (T_{max} = 826 °C, t_{85} = 10 s) modelled for the simulation of microstructural evolution of S355, based on the *Gkatzogiannis Approach* [92]

T [°C]	Microstructure					σ_{yi} [MPa]					σ_y [MPa]	α^{se} [-]
	PM	AY	FE, PE	BA	MA	PM	AY	FE, PE	BA	MA		
826	69 %	31 %	0 %	0 %	0 %	59	67	71	123	53	55	13
660	69 %	31 %	0 %	0 %	0 %	91	102	196	278	116	108	13
620	69 %	25 %	6 %	0 %	0 %	98	125	252	356	150	136	13
480	69 %	2 %	6 %	23 %	0 %	144	190	395	517	280	299	12
20	69 %	0 %	6 %	23 %	2 %	245	258	490	620	355	386	12

3.2.3 Investigated Aspects

During the simulation of components A and B for the validation of the present engineering approach several aspects of weld simulation were investigated. They are presented in Table 16. For better understanding, the full description of each investigated case is presented along with the respective results in the following chapter.

Table 16: Matrix of the simulations – The investigated concepts and the respective components

Investigated aspect	Investigated components	Results first published in	Evaluation based on
mesh convergence	A	[92]	thermal analysis
parameters of Goldak's source	A	[92]	thermal analysis
coefficient of thermal losses	A	[92]	thermal analysis
modelling of clamping mechanisms	A	[53], present	structural analysis
modelling of a restraint-free component	B	[92], [55]	structural analysis
hardening behaviour	B	[55]	structural analysis
reference temperature	B	present	structural analysis
strain rate dependency	B	[59]	structural analysis
overall validation	A, B	[92], present	thermal and structural analysis

3.2.4 Analyses and Results

3.2.4.1 Thermal Analysis

The results of the thermal analysis were taken into consideration for the evaluation of required mesh density, assignment of dimensions to Goldak's heat source and modelling of heat losses.

Mesh Convergence Study

A mesh convergence analysis was carried out in [92], in order to ensure mesh-independent results. A mesh dense enough has to be applied so that the steep temperature gradients transverse to the FZ and in the HAZ can be calculated with sufficient preciseness. In order to reduce the calculation time, mesh size away from the FZ and HAZ, where temperature changes are not so radical, was increased. Elements elongated towards the welding direction were applied. It was initially assumed that the mesh size in this direction is not critical; the heat source moves, the area to its front is melting so the thermal history up to that point is of low importance. It was ensured though, through an additional preliminary convergence analysis with smaller length of the elements, which is not presented here for the sake of space, that this option was not influencing the calculated residual stresses. An element dimension of 5 mm was then applied in all investigations of component A for the longitudinal direction. The investigated meshes had all the same pattern but the element mesh size was divided in each case with an element mesh factor, as it is presented in Table 17. Mesh *CC1* is presented exemplary in Figure 33.

Table 17: Tested mesh setups

Mesh case	Mesh factor	Mesh size in the FZ [mm]
C025	0.25	1.763
CC05	0.5	0.877
CC1	1	0.357
CC2	2	0.141
CC3	3	0.066
CC4	4	0.045

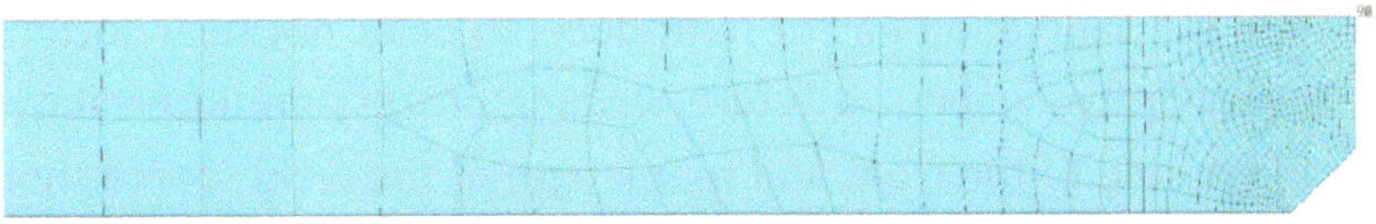

Figure 33: Pattern of applied mesh - CC1 mesh on the cross section of component A

The temperature profiles transverse to the moving heat source for the different meshes were compared in order to estimate the required mesh density. The temperature profile at the middle of the weld line, when the heat source passes by that point, were taken into consideration. As it is presented in Figure 34 mesh cases *C1* to *C4* produced identical results. Therefore, mesh size of case *C1* (0.357 mm) was applied for further simulations of component A. The calculated ratio of applied mesh size in the HAZ and FZ to the width of the Goldak's source was equal to 7 %. This normalized size of mesh was applied for all subsequent simulations, as it is rationally assumed that the required mesh density is defined by the width of the simulated heat source.

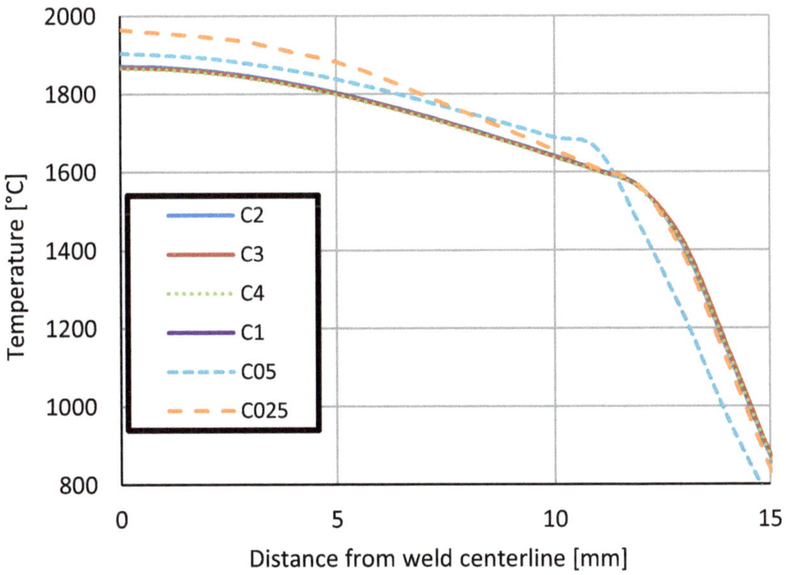

Figure 34: Results of the convergence study

Parameters of Goldak's Source

A problem that arises during the first steps of welding simulation is the assignment of proper dimensions to Goldak's heat source. The depth and width of the weld pool in an existing weld can be measured on macrosections. Nevertheless, in cases where the simulation has to reproduce an older not fully documented result, or act predictively about a future welding process an uncertainty rises. Moreover, measuring of the front and rear length of the weld source should be possible only during welding and this could be challenging. Goldak proposed in [61] that the *a* and *b* are equal to the width and depth of the weld pool respectively. He further proposed that, in absence of experimental data, the front half-length of his source should be assumed equal to one half of the width ($c_f = a$) and the rear half-length equal to twice the width ($c_r = 4a$). A set of arbitrary values for the dimensions of Goldak's source, all selected though in the same order of magnitude with the theoretical size of the weld pool, was tested on component A. The investigated cases along with the respective values for the parameters of Goldak's source, which were considered in [92], are presented in Table 18.

Table 18: Arbitrary selected and tested parameters for the Goldak's source in [92]

Cases	c_f [mm]	c_r [mm]	a [mm]	b [mm]
N	7.5	15	7.5	5
P	7.5	15	7.5	7.5
Q	10	20	10	10
R	5	10	5	3
T	5	20	5	7

The results were compared with thermocouple measurements, which were carried out by Andersson (found in [3]). The location of the three thermocouples, as it was found in [3], is presented on the sketch of Figure 35. The temperature profiles at points A, B and C are presented in Figure 36, Figure 37 and Figure 38 respectively.

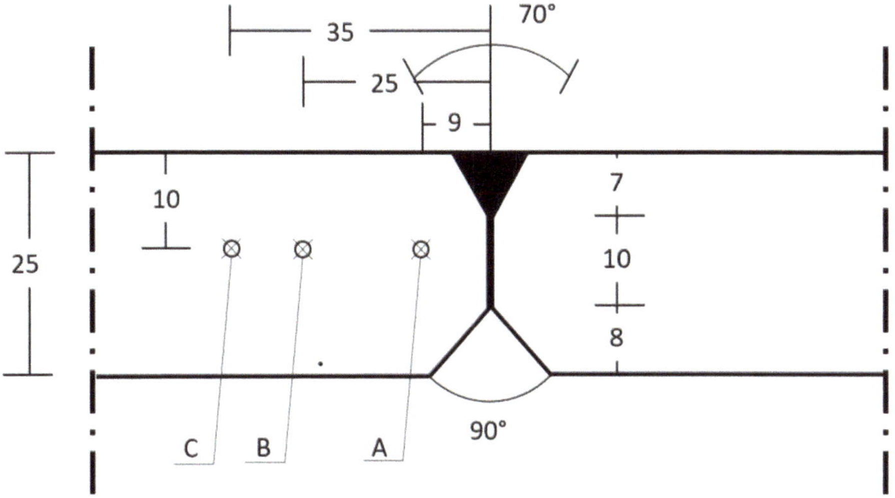

Figure 35: Location of the thermocouples A, B and C[5], dimensions are given in mm

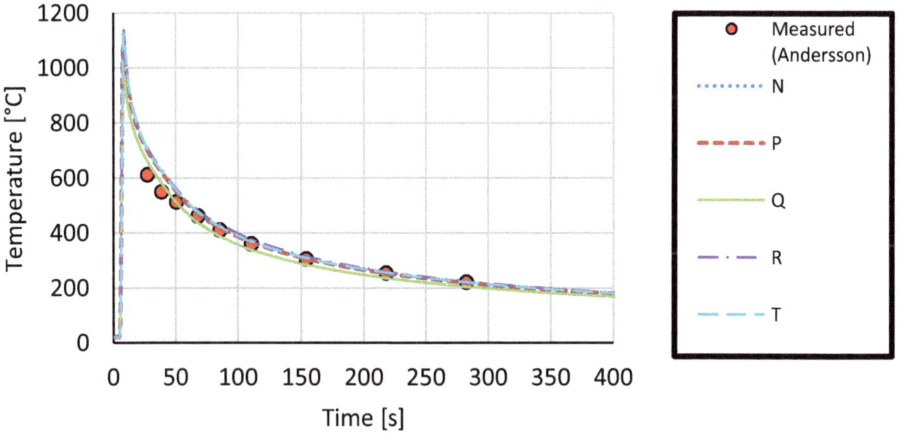

Figure 36: Dimensioning heat source – Simulated and measured temperature history at point A

All models apart from *Q* produce similar results. Accuracy of *Q* case is better in the two points near the weld, but the respective results at point C dictate the exclusion of *Q* dimensions from consecutive analyses. Apart from that, it can be concluded that small variation of the weld heat source parameters (up to 50 %) has only a limited influence on the calculated temperature history. This validates the dimensioning of Goldak's source based on the theoretical width and depth of

[5] the thermocouples were placed at a depth of 10 mm by drilling holes to the component, see [3]

the weld pool, as small discrepancies in reality from the predicted size would cause negligible influence on the thermal results.

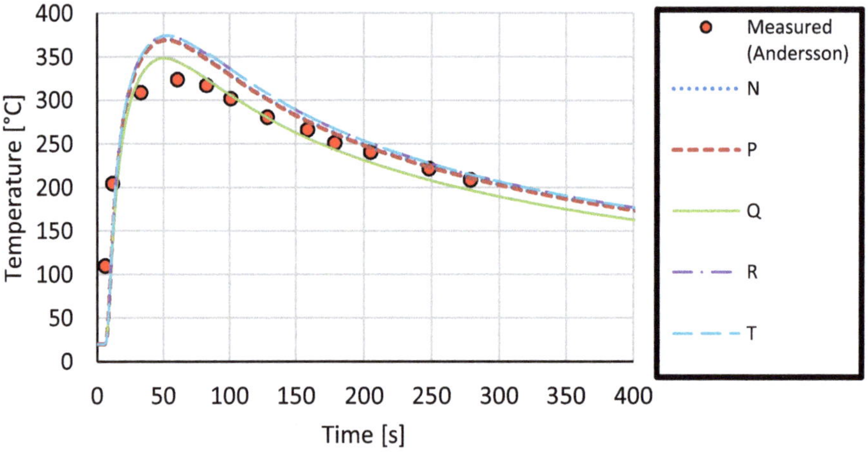

Figure 37: Dimensioning heat source – Simulated and measured temperature history at point B

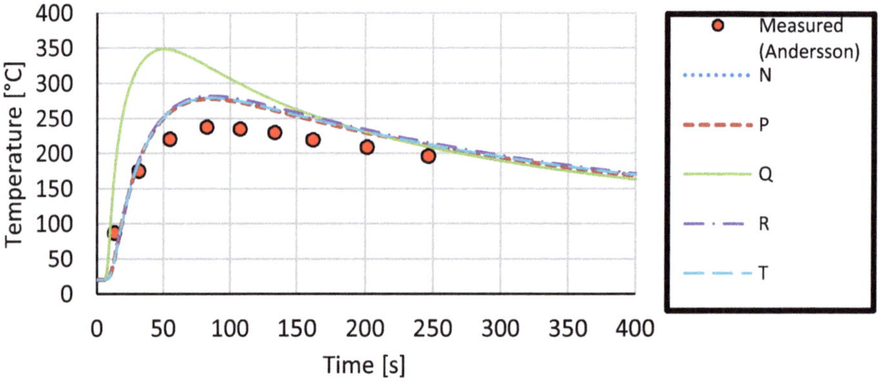

Figure 38: Dimensioning heat source – Simulated and measured temperature history at point C

Modelling of Thermal Losses

Three models with different considerations of thermal losses were applied. In case *BC1*, similar boundary conditions as those applied in [3] were used. The area around the weld source was considered thermally insulated for the first 90 s. Two different, temperature-independent coefficients were proposed for convection and radiation heat losses respectively by Anderson. Still, as they are temperature-independent, even a rough estimation with the provided values of h_C and h_R exhibits the predominance of the convective coefficient. Hence, h_R was neglected for *BC1*. Moreover, the value of 90 % proposed by Andersson for the weld arc coefficient is higher than the value of 85 % proposed by Dupont for submerged arc welding. This lower value is tested in *BC2*, retaining rest of the simulation setup same as in *BC1*. In case *BC3* a temperature-dependent coefficient for convection and radiation heat losses found in [140] was applied. Rohr [140] combined two literature sources in order to propose the specific set of values for the temperature-dependent coefficient of heat losses, which takes into consideration both convection and radiation losses. The three above-mentioned cases are presented in Table 19 and the respective results are illustrated in Figure 39 - Figure 41.

Table 19: Tested boundary conditions for the transient thermal analysis

Cases	Coefficient for thermal losses	Coefficient of heat source
BC1	temperature-independent, found in [3], radiation neglected	0.90, proposed in [3]
BC2	temperature-independent, found in [3], radiation neglected	0.85, proposed in [31]
BC3	temperature-dependent, found in [140] and presented in Table 5	0.85, proposed in [31]

The results of the simulated cases were compared with the thermocouple measurements by Andersson. Cases *BC2* and *BC3* seem to fit better than *BC1* with the measurements. Nevertheless, results from these two cases do not differ with each other significantly. It can be safely concluded then that the selection of the proper weld source coefficient is more decisive for the preciseness of the thermal results, than the use of a temperature-dependent coefficient for heat losses through convection and radiation. Modelling approach of heat losses adopted in *BC3* was finally preferred instead of *BC2* for next stages of the simulation as it is closer to physical reality.

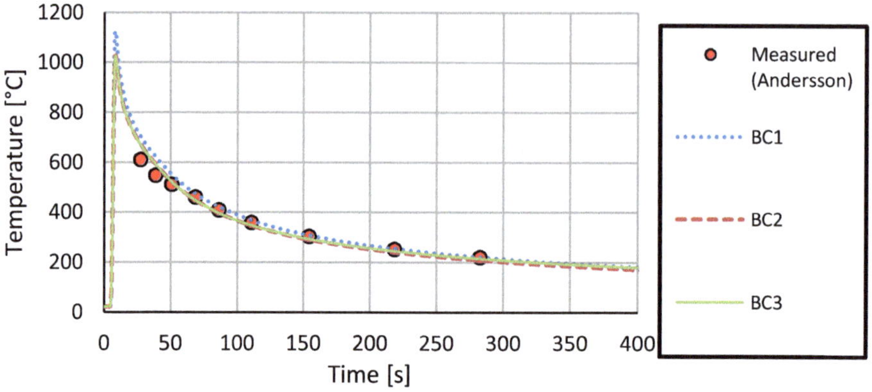

Figure 39: Modelling of thermal losses – Simulated and measured temperature history at point A

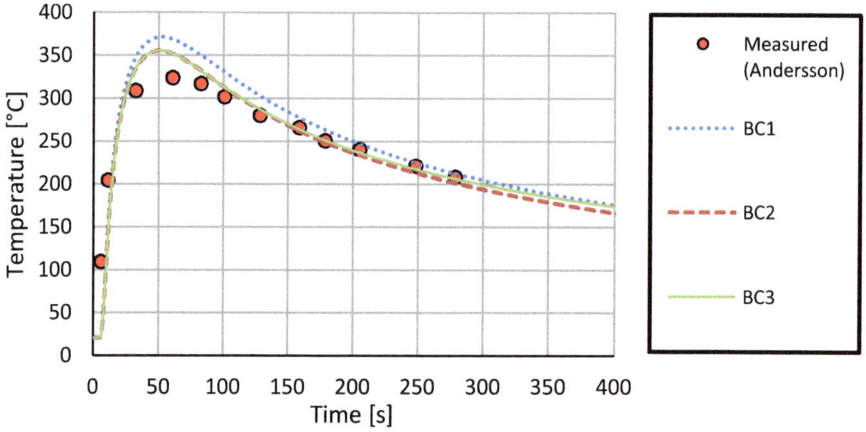

Figure 40: Modelling of thermal losses – Simulated and measured temperature history at point B

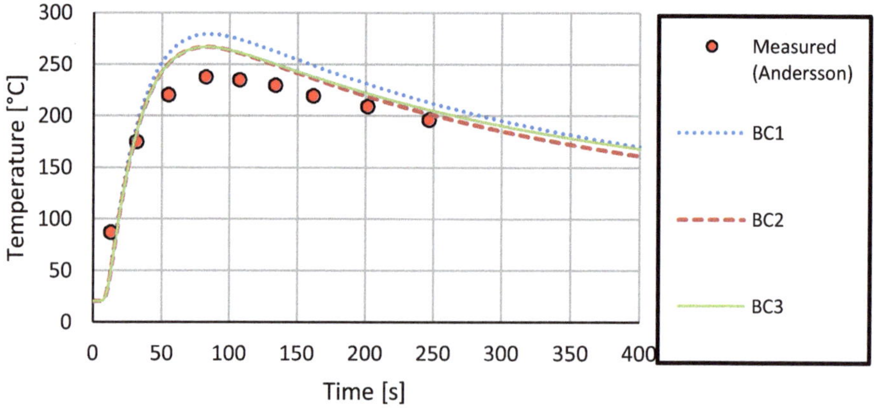

Figure 41: Modelling of thermal losses – Simulated and measured temperature history at point C

3.2.4.2 Structural Analysis

Results of the structural analysis were applied for the validation of the method regarding the calculation of WRS. Therewith, various aspects of weld simulation and their influence on the calculated WRS were investigated.

Overall Validation of the Simulated Welding Residual Stresses

Contour plots of the calculated longitudinal and transverse WRS on the top of component A, as they were published in [92], are illustrated in Figure 42 and Figure 43 respectively. Both profiles confirm the theoretical expected distribution of WRS. In the longitudinal case, tensile stresses near and inside the FZ and HAZ are met, which are at the same order of magnitude as the local yield strength. Away from the weld, counterbalancing compressive WRS are met as expected.

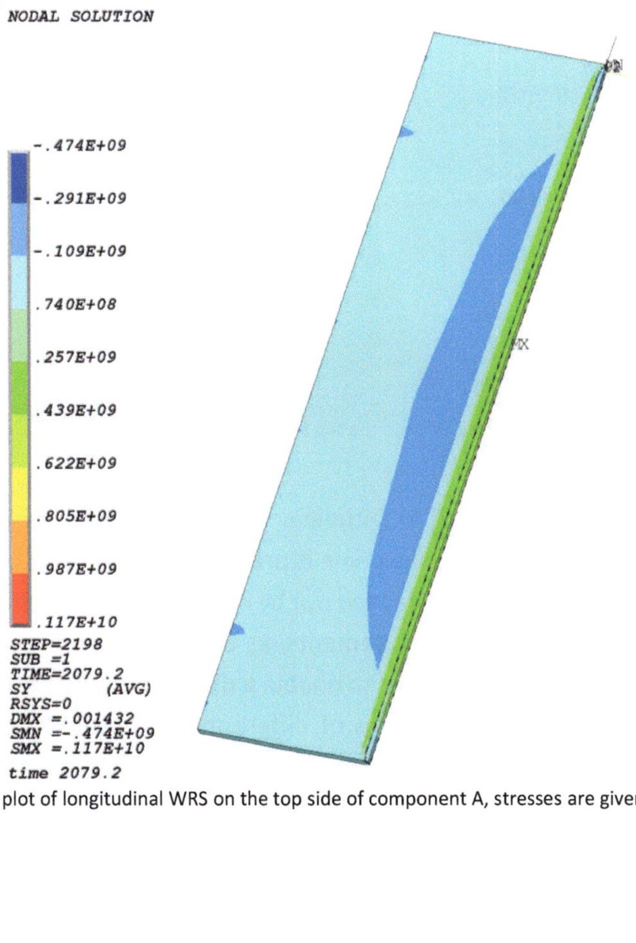

Figure 42: Contour plot of longitudinal WRS on the top side of component A, stresses are given in Pa – Initially published in [92]

In the transverse case, tensile stresses near and inside the FZ and HAZ are met as well, but this time the width of the tensile zone adjacent to the weld line is narrower. Once again, counterbalancing compressive WRS are met away from the weld.

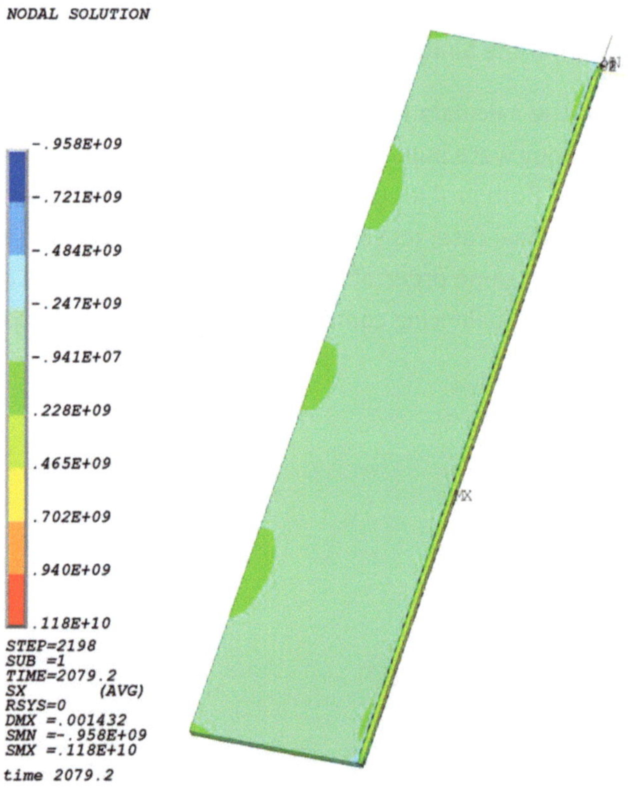

Figure 43: Contour plot of transverse WRS on the top side of component A, stresses are given in Pa – Initially published in [92]

The simulated distribution of longitudinal WRS at the centre of the top side of component A adjacent to the weld line is presented in Figure 44, along with previous numerical investigations carried out by Lindgren [111] for the same component. The respective WRS measurements, as they were provided by Andersson [3], are presented as well in order to enable a direct comparison. On the bottom of the diagram the resulting zones of differentiated material behaviour, after CMM assignment took place with the applied FE algorithm, are presented. Two different profiles of WRS, which were both calculated in the framework of the present dissertation, are presented in the diagram. The one tagged as *"Gkatzogi-*

annis, 2017" is the older one and it was published in [92]. The "*present*" calculation is an updated solution based on the previous model, but with improved material modelling, published in the present dissertation for the first time.

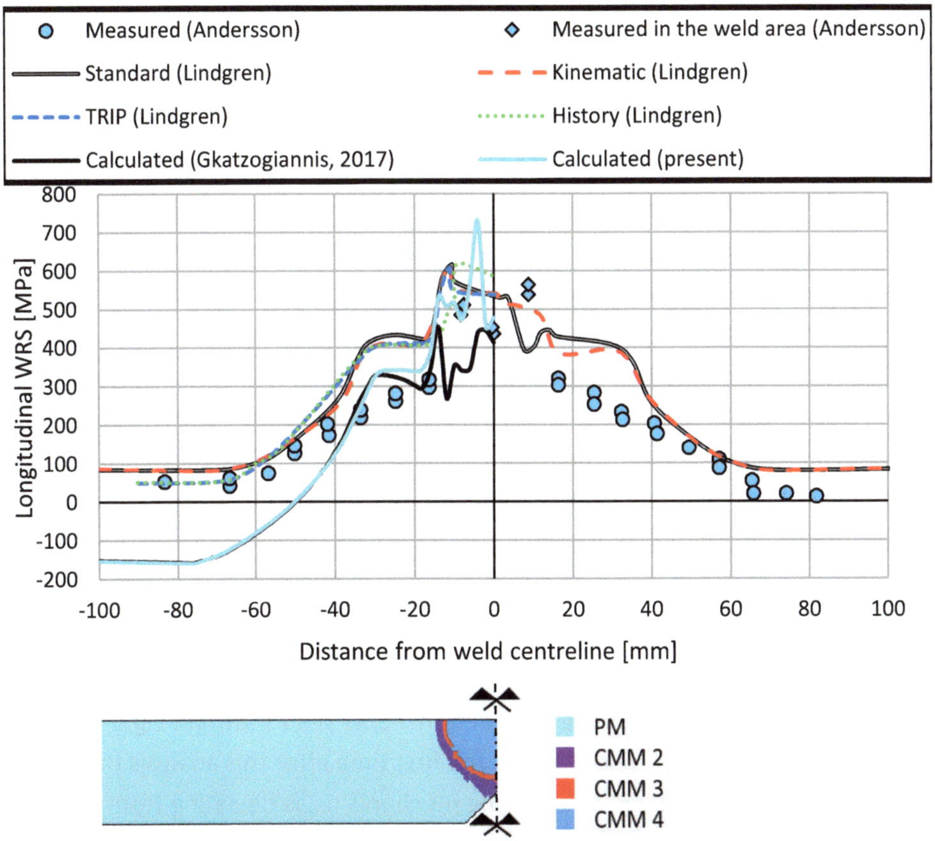

Figure 44: Validation of longitudinal WRS at the centre of the top side of component A adjacent to weld line – Measurements found in [3] – The cross section of the component at its centre transverse to the weld line along with the assigned CMMs and the symmetry condition are illustrated at the bottom

Even from the first solution though, a significantly better agreement between measured and calculated WRS close to the weld area was observed in comparison to earlier studies by Lindgren [111]. On the other hand, the previous models predicted better the WRS away from the HAZ. Nonetheless, the WRS in the HAZ are of greatest interest for the subject of the present study. This area is susceptible to fatigue loading and is the one that is treated by the HFMI treatment. Still, even from the earlier solution *(Gkatzogiannis, 2017)* [92], the predominance of the present approach was obvious. After the refinement of material modelling, the agreement of the model was further improved, and a very good agreement is met on the first 40 mm away from the weld centre, both in the HAZ and the FZ. A peak of WRS higher than 700 MPa and much higher than the nominal yield strength of the investigated material is met in the FZ. It is attributed to the changed microstructure in the respective area. Due to rapid cooling, the new microstructure has increased martensite percentage and thus higher yield strength.

The calculated transverse WRS on the top and bottom sides of component A along with the respective measurements by Andersson are presented in Figure 45. The agreement of the simulated WRS in this case was not as good as for the longitudinal. The deviation between measured and simulated WRS was significant in the weld area. Questions arise though, regarding the measurements inside the FZ. Neither special information is provided by Andersson on the technique applied and if special precautions were taken in order to take into consideration the altered microstructure, nor the measurements validate the theoretical profiles of WRS inside the weld area (Figure 3) as WRS near zero were measured. On the contrary the simulated WRS validate the theoretical ones, being tensile and in the order of magnitude of the local yield strength.

Nevertheless, agreement between measured and simulated WRS was not satisfying away from the weld, both on top and bottom of the plate, even with the updated material model. An assumption was stated at this point, by considering the influence of restraints on theoretical transverse WRS shown in Figure 3, originally found in [96]. It was regarded for the first time after this analysis that fixing the FE nodes for simulating the clamping mechanisms, is deviating from physical reality and does not allow a precise simulation of the transverse WRS. In any case, a less stiff restraint allowing in-plane small displacements should be considered instead. Much better agreement was achieved with the previous consideration though, when the initial model was resolved, by considering this time non-infinitesimal strains (large strains included), a solution that was as well published in [92]. The significant improvement of the solution, when this effect is taken into consideration, is attributed to the influence of the plate bending on the transverse WRS.

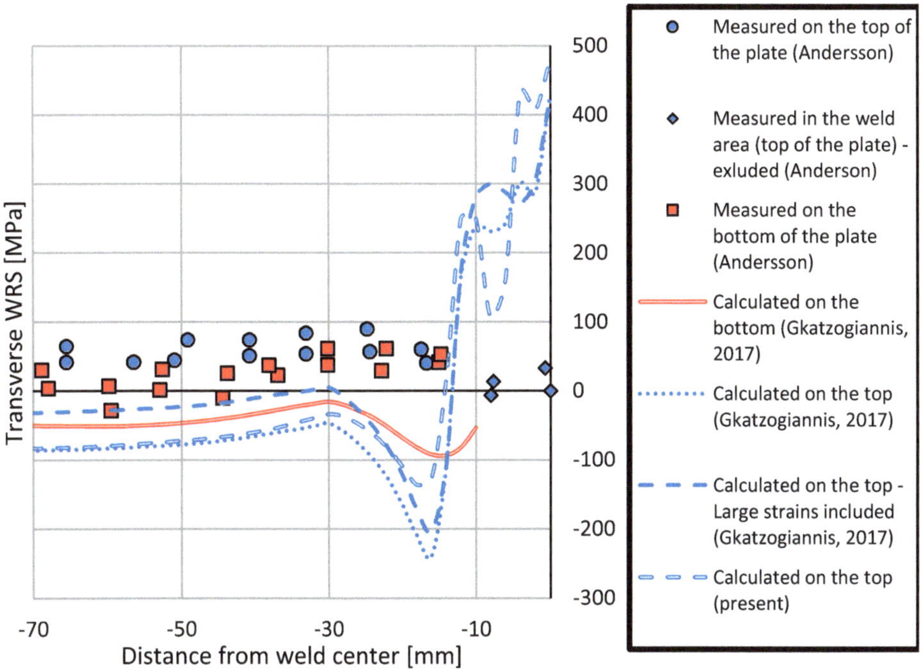

Figure 45: Validation of transverse WRS at the centre of the top side of component A adjacent to weld line – (Gkatzogiannis, 2017) refers to [92] – Measurements found in [3]

Modelling of Clamping Mechanisms

The above-stated problem of simulating clamping mechanisms, which are not fully fixed supports but allow only small in-plane displacements was tackled by a series of consecutive simulations, which were first presented in [55]. *Present* material modelling was applied for these more recent simulations. The two setups of linear spring elements of Figure 14 were tested for modelling the clamping mechanisms of component A. In the first one "*B*" (see Figure 14a) the transverse displacement of each node on the edge of the clamped areas was restrained by a spring (equal stiffness *K* for all springs). For the applied mesh density, 80 nodes were restrained in each clamping area (4 clampers, 320 springs in total along the restrained edge). The longitudinal displacement of the edge was restrained by four springs on its top and bottom corners. In all investigated cases, the total stiffness of these four springs was equal to the total stiffness of the springs, which restrained the transverse displacement so that the displacements in both directions were restrained with the same total stiffness. The vertical displacement of the nodes on the top and bottom surface of the clamped areas was fixed. With this setup of boundary conditions, small in-plane displacement of the plate in the clamped areas are possible, depending on the stiffness of the springs.

In the second proposed approach *"B2"*, only transverse displacement of the edge was restrained by springs. This simplification was based on the above-mentioned negligible influence of modelling the clamping mechanisms on the calculated longitudinal WRS. This time, both the vertical and longitudinal displacement of the nodes on the upper and bottom surface of the clamped areas was fixed (see Figure 14b). Different values of stiffness were tested for both setups. They are presented in Table 20. A more recent analysis *"Gkatzogiannis, 2019"*, identical to the previously investigated case *B21* with large strains being taken into consideration though, is presented for the first time in the present dissertation.

Table 20: Investigated cases of *stiff springs* in [53]

Cases	K [N/mm]	Spring setup
B1	10^3	B
B2	10^6	B
B3	10^9	B
B21	10^3	B2
B22	10^6	B2

The results of this series of simulations are presented in Figure 46 and Figure 47. As expected the influence on the longitudinal WRS is negligible. Cases *B2* and *B3* along with the case, where the nodes in the clamped areas were fixed, provided slightly better agreement with the measured residual stresses, with this difference though, lying inside the confidence boundaries of the present approach.

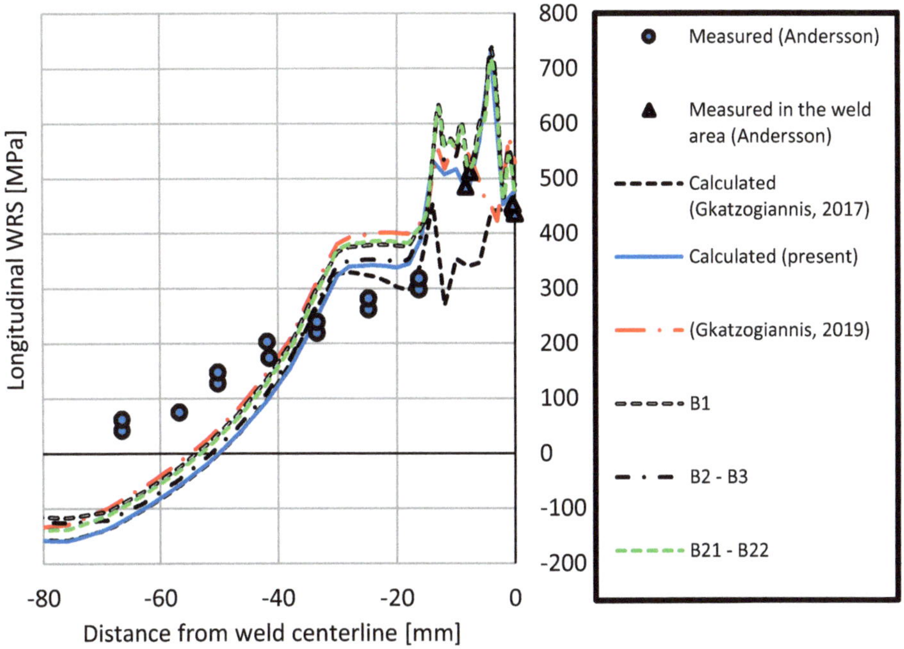

Figure 46: Longitudinal WRS at the centre of the top side of component A adjacent to weld line – (Gkatzogiannis, 2017) refers to [92] – Influence of boundary conditions

On the other hand, a significant improvement is met regarding the transverse WRS, especially with the second setup of *"stiff springs"*. The calculated profiles of *B21* and *B22* lie quite close with the measurements outside the weld area. In the case *Gkatzogiannis, 2019* , where both the concept of *stiff springs* is applied and large strains are included in the simulation, the agreement between measured and calculated transverse WRS is very good outside the weld area. The deviation between measured and calculated stresses inside the weld section in all cases is subject to the above-mentioned uncertainty regarding the profile of the measured stresses.

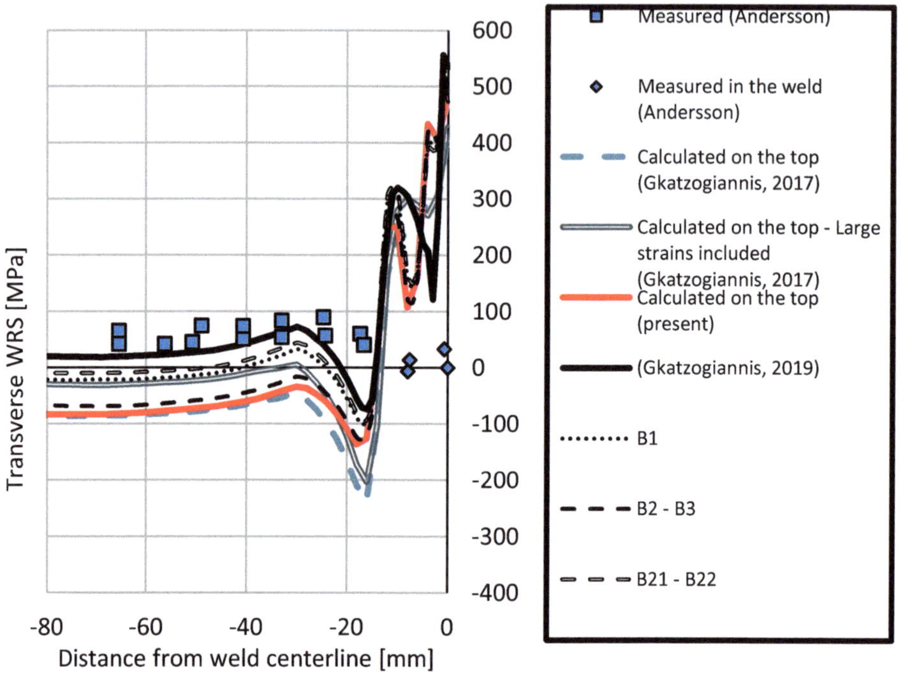

Figure 47: Transverse WRS at the centre of the top side of component A adjacent to weld line – (Gkatzogiannis, 2017) refers to [92] – Influence of boundary conditions

Modelling of a Restraint-Free Component

Simulation of the freely welded component B was achieved with the use of "*soft springs*", meaning linear spring elements subject to Hooke's law with relatively low stiffness, in comparison to the stiffness of the investigated component. The use of such boundary conditions was necessary, as the simulation of a restraint-free component would lead to numerical problems. At the same time, the applied conditions should allow the plate to deform similarly to the real component. Two different values of stiffness were tested in [53] and they are presented in Table 18. The applied set up of springs is illustrated in Figure 13.

Table 21: Applied values of stiffness *K* for the simulation of *soft springs* in [53]

Cases	*K* [N/mm]
A1	10^3
A2	10^0

The results from these analyses are illustrated in Figure 48 and Figure 49. Both cases *A1* and *A2* produce identical WRS results. It is concluded that the stiffness of 10^3 N/mm is low enough to simulate a free component. In Figure 48 the calculated longitudinal WRS with the present method produce a profile that provides results of similar preciseness with the simulation carried out by Caron et al. [21] with the commercial software *SYSWELD* [158]. In detail, each of both simulated profiles lies closer to the measured stresses at different points. The fact that the present approach follows at least qualitatively the measured stresses near the weld was encouraging for the preciseness of the present apporach, in comparison to the simulation presented by Caron et al. [21]. Similar conclusions are drawn, when the respective results for the transverse WRS are compared. Still, improvement of this model was needed at that point.

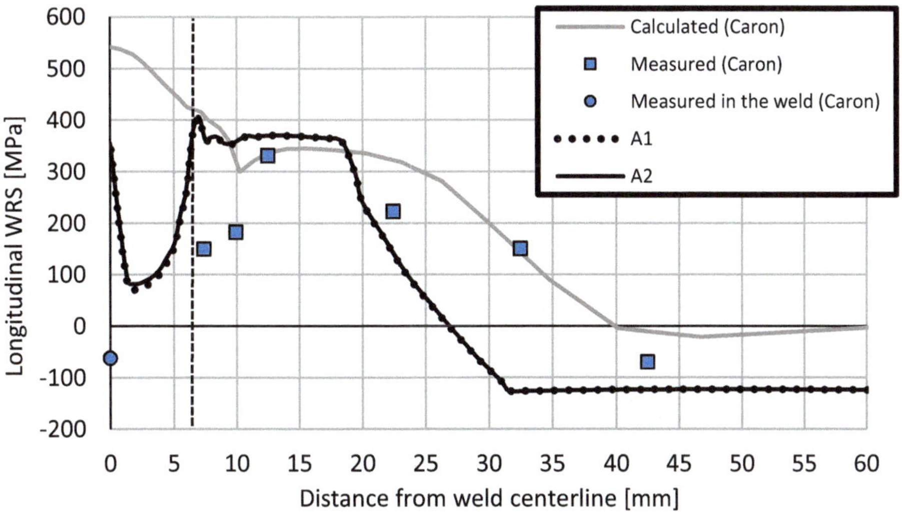

Figure 48: Longitudinal WRS at the centre of the top side of component B adjacent to weld line – Measurements found in [21] – Simulation of restraint-free conditions

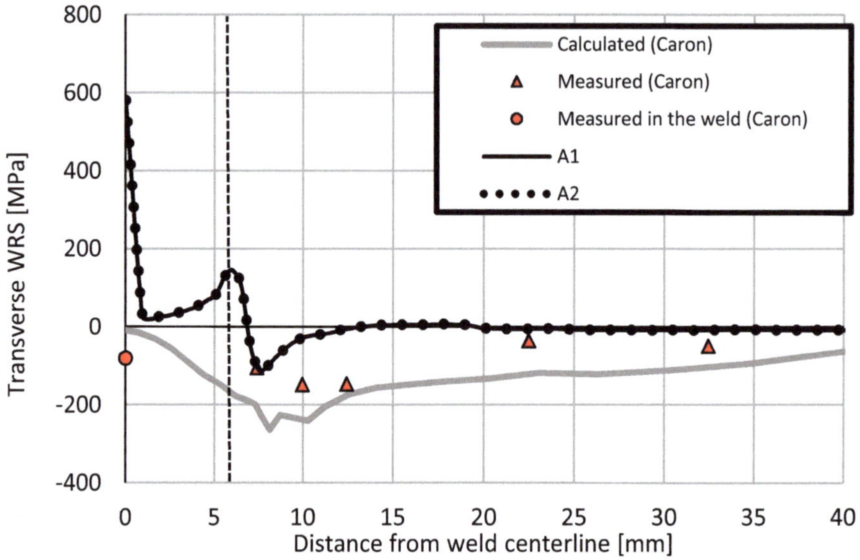

Figure 49: Transverse WRS at the centre of the top side of component B adjacent to weld line – Measurements found in [21] – Simulation of restraint-free conditions

Hardening Behaviour

The influence of the applied hardening model was investigated as well in [55]. It had already been proven in the past that an isotropic hardening behaviour is more appropriate for weld simulations of austenitic steels: In their case the Bauschinger effect is assumed to be eliminated, when the material yields at elevated temperature or is heat-treated before the reversal of loading (see [178], [126]). Both isotropic and kinematic hardening were tested on component B. It was expected that the influence of the hardening behaviour on a single-pass case would be less profound than in multi-pass welding, where consecutive cycles of reversed plasticity take place.

Indeed the difference between the two simulated profiles of longitudinal WRS is small. In agreement with previous studies, kinematic hardening underestimates the WRS in comparison with isotropic. Against initial expectations, simulation considering kinematic hardening provided better agreement than the one with isotropic hardening. Nonetheless, the deviation between the two methods was small and lies inside the boundaries of both the simulation and the WRS measurements.

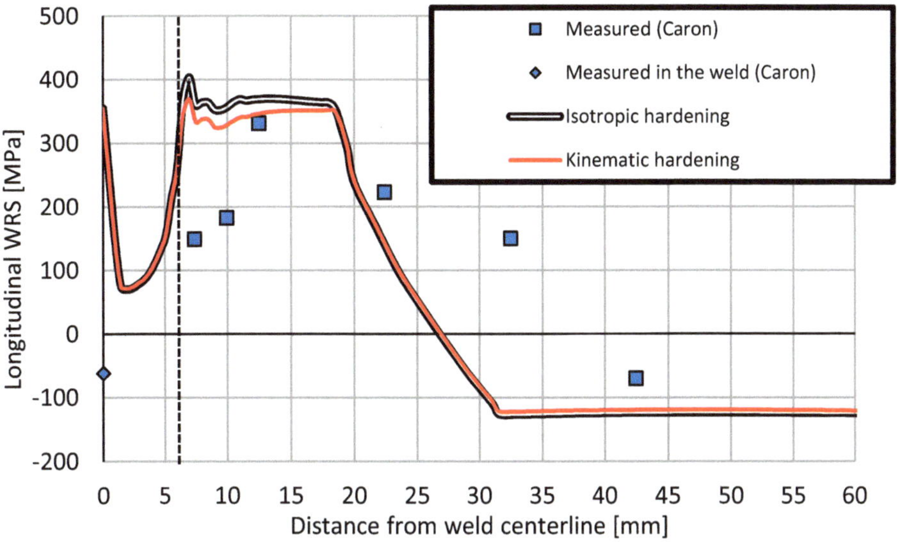

Figure 50: Longitudinal WRS at the centre of the top side of component A adjacent to weld line – Measurements found in [21] – Influence of hardening behaviour

Reference Temperature of Resolidified Material

Up to this point, all previous investigations of the author were assuming that during cooling the melted material was becoming stressed under tension from the first moment after solidification. Due to the inactivation of the respective elements above T_{melt} previous strain history was erased. As most engineering principles and laws, thermal strains are a convention describing the physical reality based on some reference. The concept of T_{ref} is easily understood from Equation (18); thermal strains depend on the reference temperature, which is selected by the modeller. Main interest of the welding simulation lies on the WRS at room temperature, thus setting the reference temperature equal with the ambient is expected. Nevertheless, this setting leads to the above-mentioned activation of elements under tension and therefore compressive plastic strains are directly introduced. This deviates from physical reality, as the resolidified material is virgin in the sense of plasticity. During its solidification, it should be stress-free. From that point on, tensile plastic strains are introduced during cooling, when the hot material shrinks and is restrained by the adjacent material. Setting the reference temperature T_{ref} of the melted material equal to T_{melt} is allowing for modelling stress-free resolidified material.

The influence of setting $T_{ref} = T_{melt}$ (case "T_{ref}") and resetting $T_{ref} = T_{ambient}$ (case "T_{ref} *(reversed)*") after cooling is investigated at this case and the results are presented in the present dissertation for the first time according to authors knowledge. The calculated longitudinal and transverse profiles of WRS are presented in Figure 51 and Figure 52 respectively. An improvement to the preciseness of the calculated WRS is met in both cases in comparison to the case, where this influence was neglected. A significant deviation between the two cases for the WRS calculated in the weld section is met, as tensile and compressive stresses are calculated in the former and in the latter one respectively. The above-described uncertainty regarding measurements in the FZ did not allow a selection of the valid modelling technique based on this deviation. Regarding the rest of the measurement points, T_{ref} and T_{ref} *(reversed)* show better overall agreement than each other in the case of transverse and longitudinal WRS respectively. Consequently, a direct interpretation at that point was not possible without further analysis. Still, it is clear that the influence of T_{ref} cannot be neglected. Therewith, the above-stated, required improvement of the model for the component B was achieved.

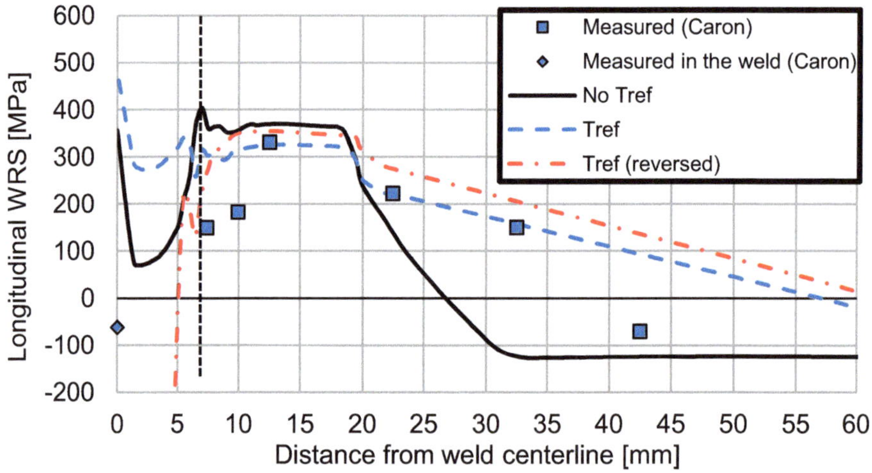

Figure 51: Longitudinal WRS at the centre of the top side of component B adjacent to weld line – Measurements found in [21] – Influence of T_{ref}

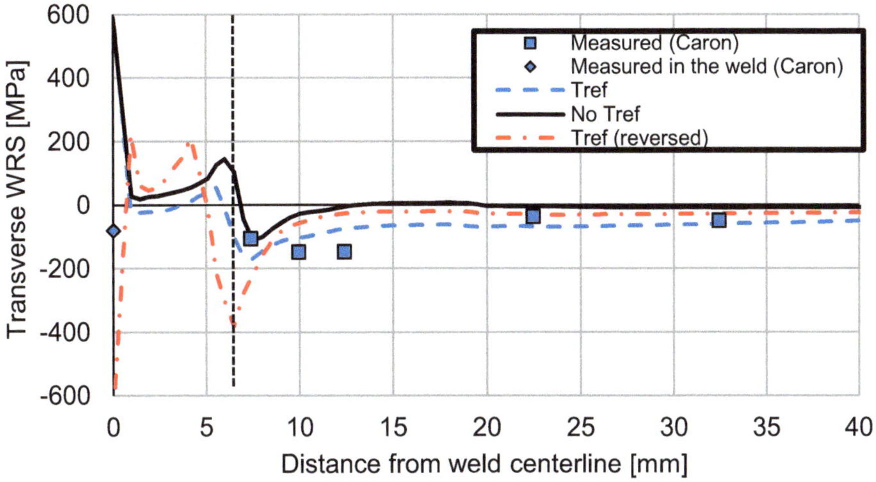

Figure 52: Transverse WRS at the centre of the top side of component B adjacent to weld line – Measurements found in [21] – Influence of T_{ref}

Strain Rate Dependency

Two different cases, *VA_TM1* and *VA_TM2* were investigated in [59] regarding the yield strength strain rate dependency of S355. Both were based on and compared with the previously modelled case T_{ref} *(reversed)*, where rate-independent plasticity only was taken into consideration. T_{ref} *(Reversed)* was notated as *"BC"* in this study (Basic Case). In *VA_TM1*, values for the coefficients of the Cowper-Symonds model (Eq. (22)), which were proposed by Jones [86], were applied for a solution using the Perzyna model (Eq. (21)). This model was selected as it is incorporated in the applied FE software *ANSYS* [14] and is available for implicit solutions. The adoption of the coefficients of the Cowper-Symonds model for the Perzyna equation is valid, due to their above-mentioned equivalency.

Still, the values for the coefficients of the strain rate hardening provided by Jones in [86], are given for high-strength steels in general and thus, their validity should be checked for each specific investigated material individually. This argument has increased validity, when the above-described deviating strain-dependent behaviour of the various alloys is considered. For this reason, during the second investigated case *VA_TM2*, the Perzyna model was calibrated specifically for the investigated structural steel S355 based on experimental results at higher strain rates and higher temperatures found in literature (see [46], [47], [48], [49] and [91]).

These references report on the same batch of steel S355, allowing a direct interpretation. A calibration of the strain rate hardening coefficients took place for the case *VA_TM2* at different temperature levels, using the statistics toolbox of *MATLAB* [121]. An overview of the investigated cases is presented in Table 22. The temperature-dependent calculated coefficients of the Perzyna model for the investigated case *VA_TM2* are presented in Table 22.

Table 22: Investigated case for strain rate dependency, α and β are the coefficients of Perzyna model

Case	Plasticity	α	β	Strain rate behaviour based on	Temperature dependency
BC	Rate independent	-	-	-	-
VA_TM1	Perzyna	3200	0.2	[86]	-
VA_TM2	Perzyna	18.540	0.2	[46], [47], [48], [49] and [91]	included

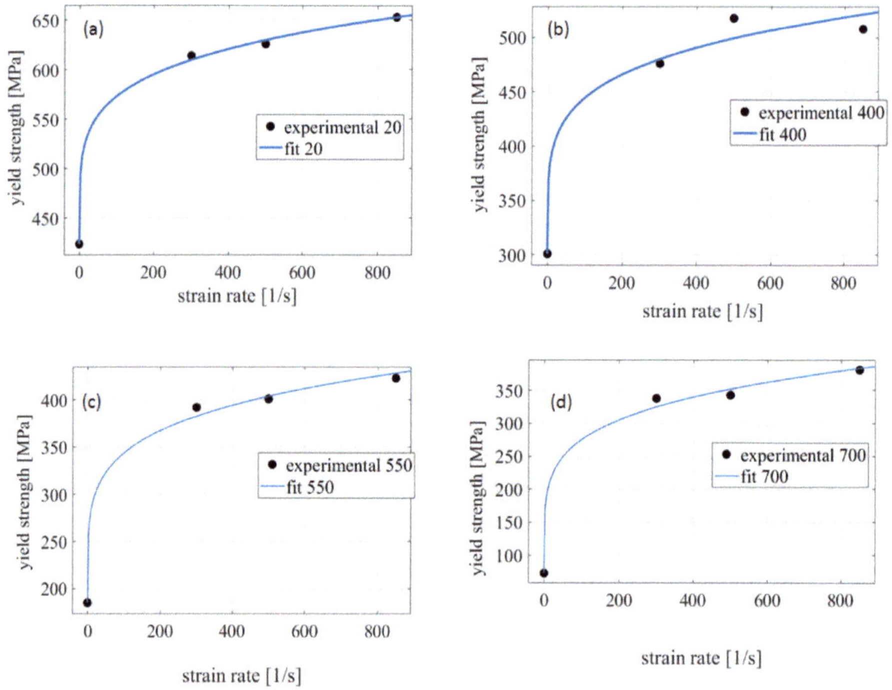

Figure 53: Fitting of the Perzyna model to the experimental data found in [46], [47], [48], [49] and [91] with *MATLAB*: a) At 20 °C; b) At 400 °C; c) At 550 °C; d) At 700 °C

The results of the simulations *BC, VA_TM1* and *VA_TM2* are presented in Figure 54 and Figure 55. The strain rate dependent models produce almost identical results, both for longitudinal and transverse WRS. A difference is met as well in both cases in comparison to the strain rate independent case *BC*. For the longitudinal stresses, both strain rate independent and dependent deviate equally from the measured WRS. In the case of the transverse stresses however, the strain rate independent case shows better agreement with the measurements. The fact that the case T_{ref} *(Reversed)* was considered as *BC* for the present analysis, while up to that point it was not clear if this would be the most suitable consideration of T_{ref} for subsequent analyses, can be accounted for this incompatibility. Nonetheless, the non-negligible difference of the results, when strain rate dependency is taken into account, has to be highlighted at this point. Although, in the current case slightly worse agreement is met, the non-negligible deviation of results shows that when high accuracy is required, strain rate dependency must be considered as it was proposed by Lindgren [111].

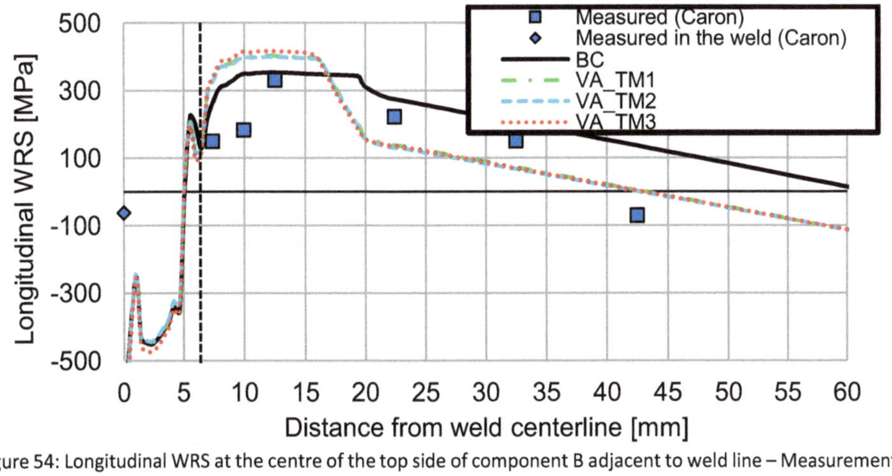

Figure 54: Longitudinal WRS at the centre of the top side of component B adjacent to weld line – Measurements found in [21] – Strain rate dependency of simulated WRS

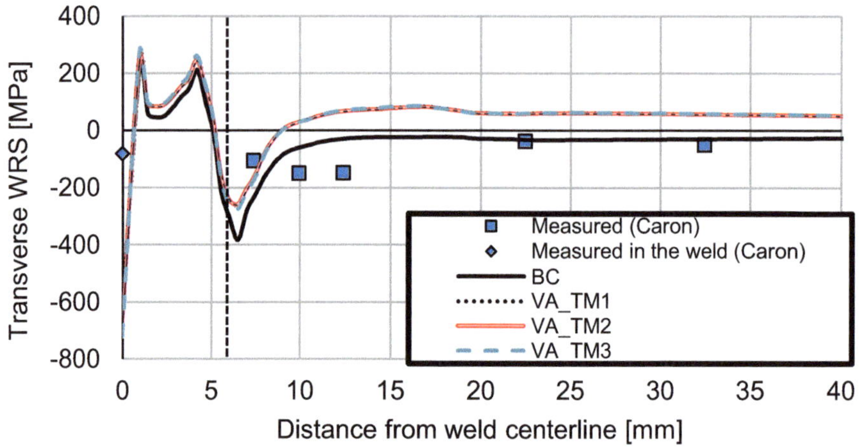

Figure 55: Transverse WRS at the centre of the top side of component B adjacent to weld line – Measurements found in [21] – Strain rate dependency of simulated WRS

3.2.5 Conclusions

A series of analyses was carried out regarding single-pass butt welds. This weldment was selected, due to its simplicity for a first-step overall validation of the presented engineering approach and for the investigation of several practical aspects of weld simulation by excluding the influence of more complex geometry or multiple weld passes.

The following conclusions were drawn regarding the overall evaluation of the presented approach:

- The current approach, additionally to its straightforwardness, provides results with preciseness within the required boundaries that were set at the beginning of the present study. Moreover, it shows better agreement with respective WRS measurements in the FZ and HAZ, which are critical for the fatigue behaviour of the component, when compared with other numerical simulations found in literature.
- The approach is straightforward. Sophisticated aspects of material science like phase transformations or TRIP, are taken into consideration, based on practical simplifications, without changing the FE formulations. The phase changes are predicted based on existing CCT diagrams and no numerical solution is required. Its straightforwardness constitutes the approach reproducible in practice by a structural engineer with only basic knowledge of

material science. The approach enables the calculation of WRS and in extension the estimation of the fatigue behaviour of a component, allowing for an adjustment in each investigated case of the preciseness and computational effort by increasing or decreasing the number of CMMs, which are modelled.

- Although the approach was originally developed to act predictively, it can also be applied for existing weldments, based on destructive testing. Hardness measurements on a macro-section can provide enough information about the yield limit of each area of the weld. Therefore, alternatively to the CCT diagrams the hardness measurements can provide information necessary for modelling the mechanical behaviour during cooling.
- The practical nature of the method allows its application on other materials as well, as it was shown for aluminium in [43], whereby different microstructural effects must be modelled. Appropriate assumptions also allow extension of the model to multi-pass welding, as it is discussed below.
- The present method allows for flexibility. The level of preciseness depends on the number of cooling-down material models and can be selected by the modeller. Depending upon the case, further, more sophisticated modelling of mechanical behaviour can be deployed, like mixed-hardening constitutive laws or strain rate dependent plasticity. The computational effort increases at the same time.

The following conclusions were drawn regarding specific aspects of the simulation:

- In order to provide mesh independent results, a mesh dimension in the HAZ and FZ transverse to the weld line equal or smaller than 7 % of the Goldak's source half width is required, when first order hexahedral finite elements are applied.
- When no relevant information is available, setting the width a and depth b of Goldak's source equal to the theoretical width and depth of the weld seam provides sufficient results. In this case, forward and rear lengths are set as $c_f = 4 \cdot a$ and $c_r = a$.
- Regarding the preciseness of thermal analysis, the selection of a proper value for the weld source coefficient is more decisive than the use of a tem-

perature-dependent coefficient for heat losses through convection and radiation. Still, both effects should be considered for safer results, as modelling with a temperature-dependent coefficient lies closer to physical reality.

- Accurate modelling of material behaviour is predominant for the improvement of the preciseness, especially when the longitudinal WRS are considered.
- Considering large strains is predominant for the precise calculation of transverse WRS.
- When a clamped-component is simulated, modelling of boundary conditions with *stiff springs* instead of zero displacement restraints improves significantly the preciseness of transverse WRS. This deviation validates the influence of external restraint on the transverse WRS as it is presented in Figure 3. When this modelling technique is combined with consideration of large strains very precise calculation of transverse WRS can be achieved.
- B2 Setup of the *stiff springs* provided the best improvement in simulations results. Application of the springs is mesh dependent and the individual stiffness of the springs has to be adjusted depending on their number. A mesh-independent stiffness density of 20 x 10^7 (N/mm)/mm was applied in the respective areas in order to simulate each clamping mechanism.
- Application of two *soft springs* in each normal direction with stiffness of 10^3 N/mm or lower is appropriate for the simulation of a restraint-free component.
- Neglecting the *Bauschinger* effect by applying isotropic hardening behaviour leads to overestimation of the WRS in comparison to results with the kinematic hardening. Selection of the appropriate approach could not be clarified with this series of analyses.
- Setting T_{ref} of the inactivated elements in the FZ equal to T_{melt} provides results with improved preciseness. However, it is not clear from this series of analyses if T_{ref} should be reset equal to ambient after cooling.
- Strain rate dependency should be taken into consideration, when high preciseness is required.

The above-presented conclusions were used as a basis for the subsequent simulations that were carried out in the framework of the present dissertation.

3.3 Further Applications of the Proposed Approach

The present method was extended for modelling single-pass aluminium welds [93] and multi-pass butt welds of austenitic stainless steels [56].

In the first case, the same method as for single-pass welds of steel was applied. Material models were assigned to different areas of the weld by considering this time the influence of heat input on recrystallization of the aluminium microstructure instead of the phase changes. This recrystallization can cause a local reduction of the yield strength for high strength aluminium alloys like in the case of EN AW 6060 [37]. It was proven, that due to this effect the highest WRS are met on the boundaries between HAZ and PM. Due to the different thermal behaviour of aluminium this boundaries lie further away from the weld toe in comparison to steel weldments. Therewith it could be interpreted, why preliminary HFMI treatment on the weld toe of aluminium welds provided improvement of their fatigue behaviour less than expected.

In the case of multi-pass welds, austenitic steels were modelled so that influence of heat input, welding direction and welding sequence could be investigated by excluding the influence of microstructural effects. Reduction of welding speed, avoidance of intermediate cooling, and antisymmetric welding sequence were proven, according to initial expectations, beneficial for the reduction of WRS and in extension for the prolongation of the fatigue life of the investigated weldments. These conclusions could be very useful for the optimization of welding process for austenitic steels. Nevertheless, they cannot be directly adopted for the case multi-phase steels, where microstructural changes can influence profoundly the calculated WRS. The theoretical background and modelling aspects for extending the present approach in the case of multi-pass welds on multiphase steels was discussed in [57].

Detailed description of these investigations is excluded from the present dissertation as they, although being relevant to the general subject, are not directly contributing to the fulfilment of the present research goal, i.e. the simulation of RS from welding and HFMI. Further details regarding these investigations can be found in the respective, given publications.

3.4 Fillet welds

A series of numerical investigations was carried out in the framework of the present dissertation, in order to calculate the WRS of the fillet welds from the research project *HFH-Simulation*. Some of the results were carried out as part of the project as well, while other are published in the present dissertation for the first time. Goal was the precise calculation of the WRS field, which would be applied as input to the subsequent HFMI analysis. The above-mentioned conclusions drawn from the former investigations on single-pass butt welds constituted the base for this latter simulation. Measurements of WRS, which were carried out in the framework of the project *HFH-Simulation*, are used for validation of the present numerical analyses. Previously unsolved issues regarding practical aspects of modelling were as well investigated in this series of analyses.

3.4.1 Investigated Components

The investigated components of the project *HFH-Simulation* were fillet welds of three different materials, namely S355 [29], S690 and S960 [30]. They consisted of a plate with dimensions 1000 mm x 370 mm x 10 mm and transversal stiffeners with dimensions 1000 mm x 50 mm x 10 mm welded on both sides. In the case of the S355 component, length was in reality 1350 mm. Nonetheless, a length of 1000 mm was assumed in the numerical models for this case as well, in order to reduce the computational time. No influence on the calculated WRS is expected, as the length remains significantly larger than the width of the plate. The geometry of the investigated component and the weld section are illustrated in Figure 56 and Figure 57 respectively. The real component of steel S355 is presented in Figure 59. It was restrained with two clampers during welding, on the side of the 1st and 4th pass (setup of clamping mechanisms *FWBC1*). Applied welding sequence A-B-C-D is as well presented in Figure 57 . Coarse mesh was applied away from the weld, while finer mesh dimensions, with dimensions of 0.37 mm and 1 mm transverse to and along the weld direction respectively, were selected for the FZ and the HAZ based on previous investigations.

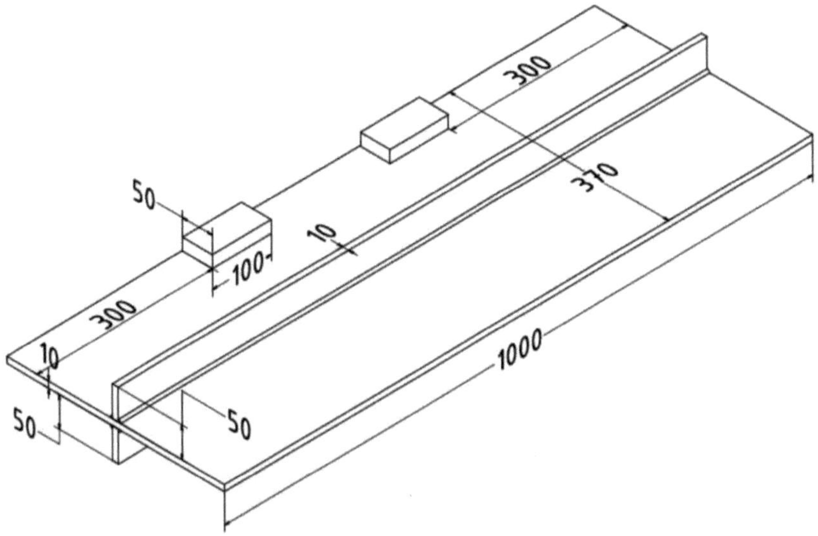

Figure 56: Fillet welds of the project *HFH-Simulation*, two clampers were applied on the left side during welding, dimensions are given in mm (setup *FWBC1*)

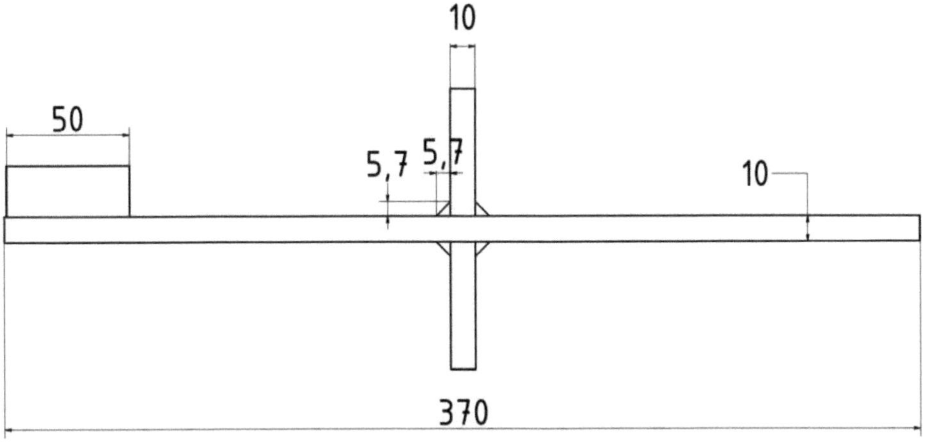

Figure 57: Component of the project *HFH-Simulation*, two clampers were applied on the left side during welding dimensions are given in mm (setup *FWBC1*)

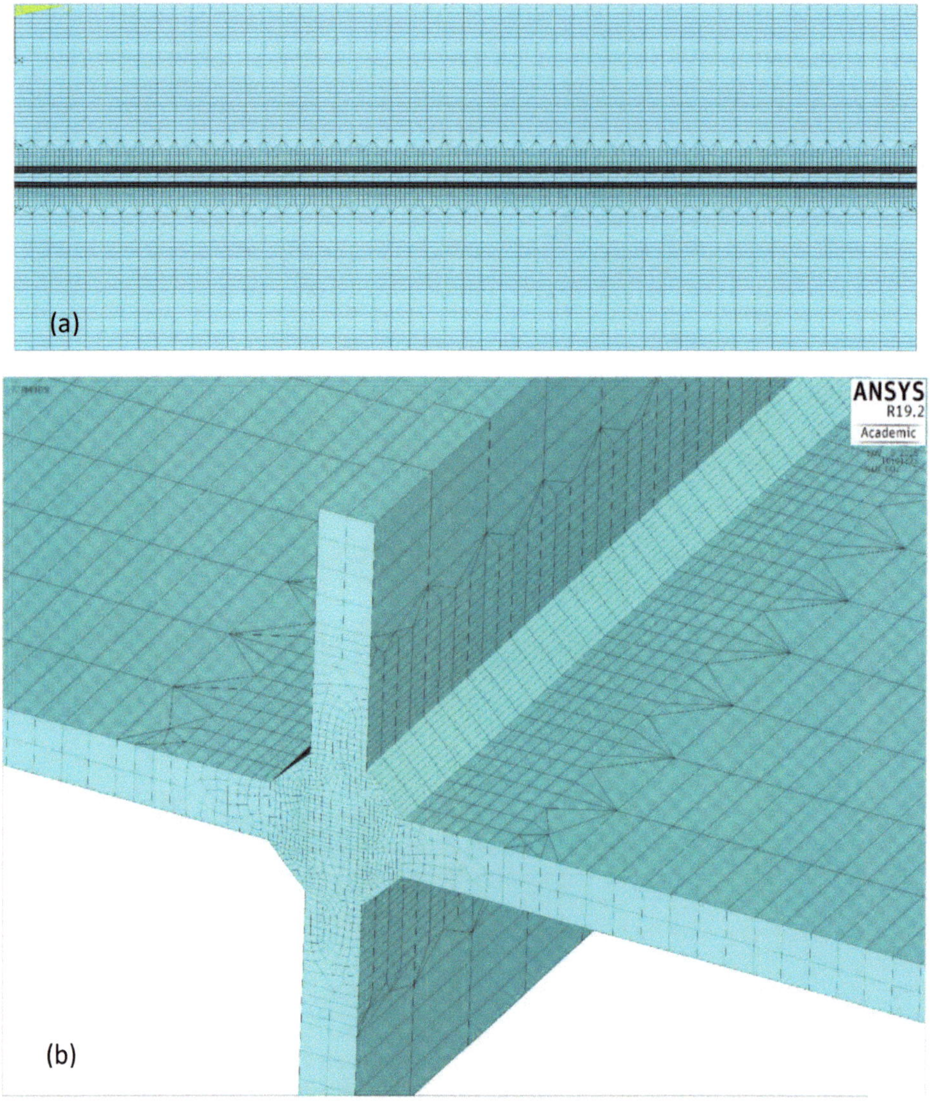

Figure 58: Mesh applied for the simulations of fillet welds: a) Top view; b) Isometric view

The component of S355, hereafter called simply FW355, was MAG-welded with mean electric power of 4.125 kW and welding speed of 6.5 mm/s (39 cm/min; 0.64 kJ/mm gross heat input). FW690 and FW960 were MAG-welded with mean electric power of 7.293 kW and welding speed of 6.5 mm/s (39 cm/min; 1.12 kJ/mm gross heat input). Thermocouples' measurements were carried out in both cases. Welding setup of FW690 is shown in Figure 60. Three stiffeners were welded on each side of a larger plate, which was later cut into the desired geometry, as the one presented in Figure 56. The same setup was applied for the FW960. Both materials were welded with automated procedure by a robot-welder, which welded simultaneously the two upper and consecutively the two bottom passes. Therefore, a second FE model was built, identical with the one presented in Figure 54 but with symmetry conditions on the middle plane of the stiffeners. Nevertheless, sequential welding sequence identical to that of FW355 was as well modelled, so that the most unfavourable case, i.e. the weld toe with the highest tensile WRS, could be taken into consideration to the subsequent simulation of HMFI as well. As it was not documented, on which plate the WRS measurements of the Project HFH-Simulation were carried out, the setup *FWBC1* was assumed for FW690 and FW960 as well. It was decided that if no satisfying agreement was achieved, further actions would be taken. WPS for all materials are presented in Appendix A. In all cases, the stiffeners were initially tack-welded on the plates, with full welding taking place afterwards. During the simulation, both the tack welding procedure and the discontinuity between the stiffener and the base plate were neglected, as they were rationally believed to have negligible influence on the final WRS.

Figure 59: The real component of the project *HFH-Simulation* made of steel S355 after completion of the welding procedure – The clamping mechanisms are seen on the left side

Figure 60: Welded plates of S690 from the project *HFH-Simulation*

3.4.2 WRS Measurements

Measurements of the WRS took place in the framework of the project *HFH-Simulation*. The measurements were carried out by the *Fraunhofer Institute for Mechanics of Materials* (IWM)[6] with an X-Ray diffractometer[7] on the 1st welding pass of each fillet weld at the middle of the plate and transverse to the welding line. In the case of S355 three measurement lines, parallel to each other at a distance of some mm from each other (≈ 10 mm), were measured. Further details regarding the measuring method can be found in [146], but they are excluded from the present dissertation as they exceed the boundaries of the present subject.

The measured WRS for all three investigated materials are presented in Figure 61 along with their confidence boundaries and the mean line for the measurements of S355. Despite the fact that all measurements were carried out on the 1st welding pass, they are presented schematically in such a way that the transverse WRS can be found on the left hand side of the diagram and the longitudinal on the right hand side. Against initial expectations, the transverse WRS range in all cases in the compressive zone and they exhibit a very strong fluctuation. A significant deviation of even up to 250 MPa is met between the parallel measurements of S355. Such a conclusion cannot be drawn for the other two steel grades due to the single measurements but a very irregular WRS-pattern with steep peaks and

---

[6] IWM stands for „Institut für Werkstoffmechanik“, German for "Institute for Mechanics of Materials"

[7] a Stresstech G3 Diffractometer was used

valleys is met especially in the case of S960. On the other hand, longitudinal measurements exhibit no significant fluctuation and the deviation for the measurements of S355 is significantly smaller. Moreover, the longitudinal profiles are closer to initial expectations, as tensile WRS are met near the weld toe that move to the compressive area away from it. Nevertheless, it seems that the whole profile is shifted downwards as the tensile and the compressive WRS have lower and higher magnitude than expected respectively and the width of the tensile area is quite narrow.

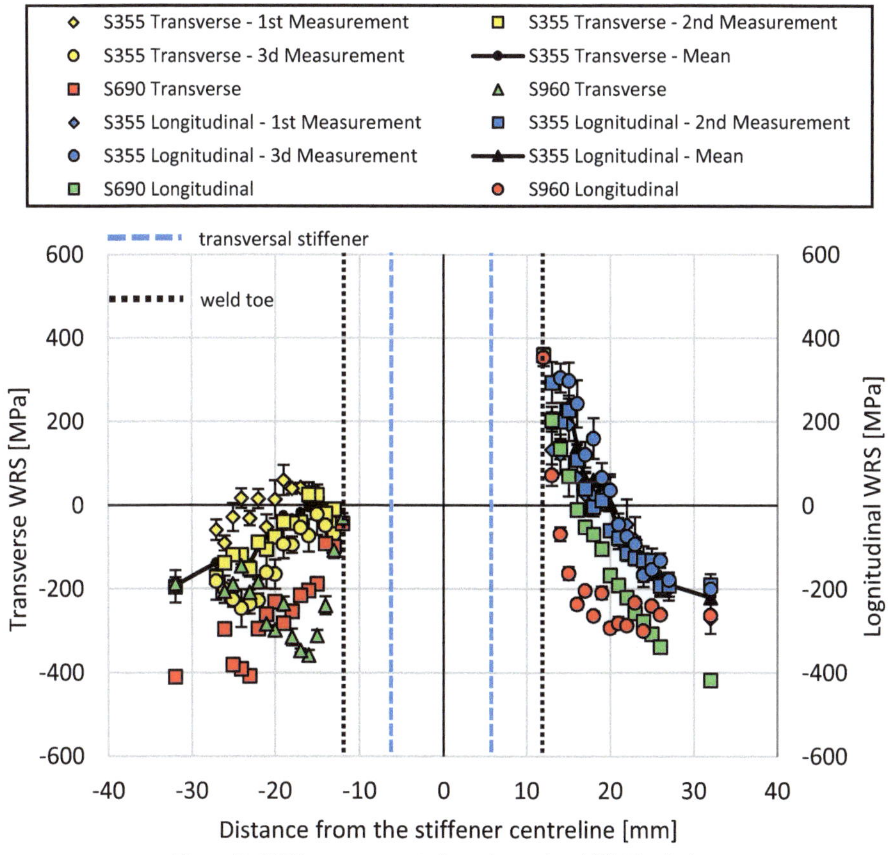

Figure 61: WRS measurements from the project *HFH-Simulation*

It was assumed that the rolling process of the parent plates or the shot blasting of the surface prior to welding should be accounted for these discrepancies of measured WRS. For instance, shot blasting can introduce significant compressive WRS (Figure 15), but also its localized mechanical effect can lead to significant differentiation of the RS, which could explain the present significant deviations and fluctuations. A similar effect, i.e. presence of compressive stresses of the magnitude of 200 MPa prior to welding was attributed to hot rolling of the investigated steel plates, in a previous numerical study of HFMI [150]. In any case, the longitudinal measurements are considered more valid due to the absence of significant fluctuations and the small deviation of the measurements for S355.

3.4.3 Material Modelling

3.4.3.1 S355

Thermal material properties given in Table 11 were applied for steel S355. Temperature-dependent mechanical properties of the investigated S355 batch were extracted in the framework of the project *HFH-Simulation*. Material characterisation was carried out from ambient up to a temperature of 560 °C. All intermediate values were interpolated, while for values above the investigated range an extrapolation was carried out, considering as well the austenitization effect. The applied values are presented in Table 23.

Table 23: Mechanical parent material properties of S355 batch used in the project *HFH-Simulation* [146]

T [°C]	σ_y [MPa]	E [GPa]	H [GPa]
20	400	233	2.80
400	300	188	4.20
500	260	162	2.30
560	240	139	2.20
719	167	55	1.84
915	45	10	1
1500	7	10	1
3000	7	10	1

The chemical composition of S355 batch and the results of the analytical calculation of the austenitization temperatures *A1*, *A3* and *A3'* are presented in Table 24 and Table 25 respectively. Based on the results of the thermal analysis, which are presented below, it was considered adequate to model 4 CMMs, as the areas in the HAZ and FZ exhibited similar t_{85}. They were built based on the CCT diagram of S355, which is presented in Figure 62, and the material properties of the individual phases (see Table 6). The CCT diagram was found in [147] and was extracted by a batch with similar chemical composition. The values of the double parameter T_{max} and t_{85} characterizing each CMM are provided in Table 26. Exemplary, evolution of phase composition and the respective mechanical behaviour of CMM 2 are provided in Table 27. For the sake of shortness, the rest of the CMMs are provided in Appendix A.

Table 24: Mean values of chemical composition for the S355 batch used in the project *HFH-Simulation* [146]

C	Si	Mn	P	S	Al-T	B-T
0.156	0.176	1.41	0.0125	0.0028	0.0265	0.00015
Cr	Cu	Mo	N	Nb	Ni	Ti
0.0305	0.013	0.0055	0.0067	0.015	0.0275	0.014

Table 25: Calculated austenitization temperatures of the S355 batch used in the project *HFH-Simulation* [146]

A1 [°C]	*A3* [°C]	*A3'* [°C]
713	815	908

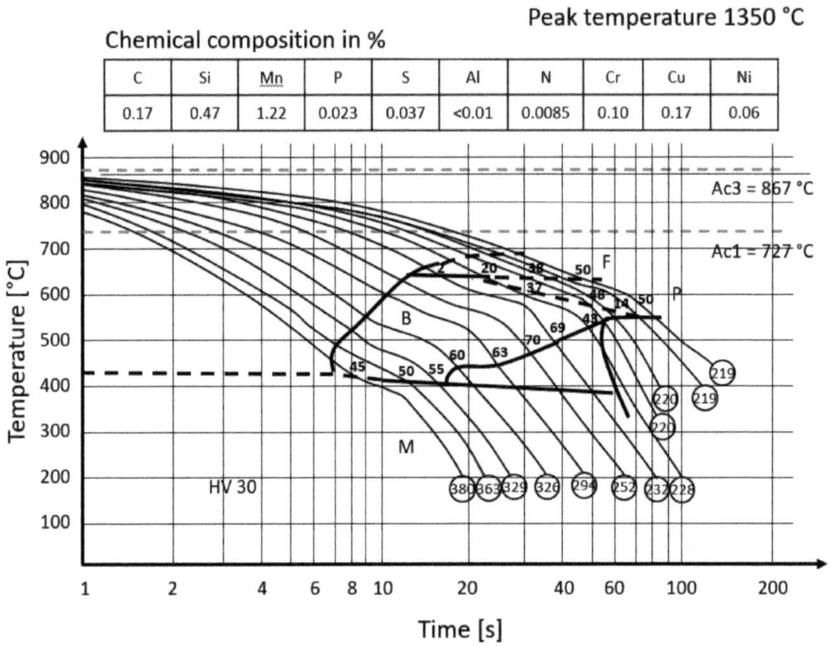

Figure 62: CCT diagram for steel S355, based on a diagram found in [147]

Table 26: Double parameter of the cooling material models for the simulation of FW355

CMM	2	3	4
T_{max} [°C]	779	844	910
t_{85} [s]	4	4	4

Table 27: CMM 2 (T_{max} = 779 °C, t_{85} = 4 s) modelled for the simulation of FW355

T	Microstructure					σ_{yi} [MPa]					σ_y	α^{se}
[°C]	PM	FE, PE	BA	MA	AY	PM	FE, EP	BA	MA	AY	[MPa]	[-]
812	66 %	0 %	0 %	0 %	34 %	100	79	42	120	62	87	13
713	66 %	0 %	0 %	0 %	34 %	153	86	73	264	81	128	13
420	66 %	0 %	0 %	0 %	34 %	292	218	237	783	136	239	13
200	66 %	0 %	0 %	31 %	3 %	353	285	271	880	200	514	12
20	66 %	0 %	0 %	31 %	3 %	400	431	277	1008	245	586	12

3.4.3.2 S690

The temperature-dependent thermal properties of S690 were found in [140] and they are presented in Table 28. Parent material mechanical properties at ambient temperature were extracted from the material characterisation of the investigated batch in the framework of the project *HFH-Simulation*. Same values of tangent modulus and analogous temperature-dependency of yield strength and Young's modulus to that of S960 was assumed. The applied parameters are presented in Table 29. The chemical composition of S690 batch, which was investigated in the project *HFH-Simulation* and the results of the analytical calculation of the austenitization temperatures *A1*, *A3* and *A3'* are presented in Table 30 and Table 31 respectively. Based on the results of the thermal analysis, which are presented below, it was considered adequate to model 4 CMMs, as the areas in the HAZ and FZ exhibited similar t_{85}. They were built based on the CCT diagram of S690, which is presented in Figure 63 and the material properties of the individual phases (see Table 6). The CCT diagram was found in [12] and had comparable chemical composition to the present. Any effect of the high heating rate of 6000 K/s of the experiments, from which the CCT diagram was extracted, is neglected in the present case. The values of the double parameter T_{max} and t_{85} characterizing each CMM are provided in Table 32. Exemplary, evolution of phase composition and the respective mechanical behaviour of CMM 3 are provided in Table 33. For the sake of shortness, the rest of the CMMs are provided in Appendix A.

Table 28: Temperature-dependent material parameters of S690 found in [140], used in thermal analysis

T [°C]	K [W / (m · K)]	T [°C]	C [J / (kg · K)]	T [°C]	ρ [kg / m³]
20	46.1	50	480	20	7880
400	44.65	200	540	200	7810
700	42.62	400	620	400	7740
1100	32.83	700	920	700	7640
1300	37.18	900	620	900	7620
1387	31.71	1200	720	1200	7470
		1381	810	1250	7440
				1300	7420

Table 29: Temperature-dependent material parameters of S690 extracted in [146], used in mechanical analysis – Values are extrapolated for temperatures higher than 560 °C

T [°C]	σ_y [MPa]	*E* [GPa]	*H* [GPa]
20	734	203	2.33
400	612	169	3.40
719	150	41	1.63
915	45	10	1
1500	7	10	1
3000	7	10	1

Table 30: Mean values of chemical composition for the S690 batch used in the project *HFH-Simulation*

C	Si	Mn	P	S	Al	V
0.14	0.4	1.38	0.007	0.002	0.03	0.061
Cr	Cu	Mo	Ti	Nb	Ni	
0.01	0.02	0.002	0.024	0.03	0.02	

Table 31: Calculated austenitization temperatures for the simulation of FW690

A1 [°C]	*A3* [°C]	*A3'* [°C]
720	835	935

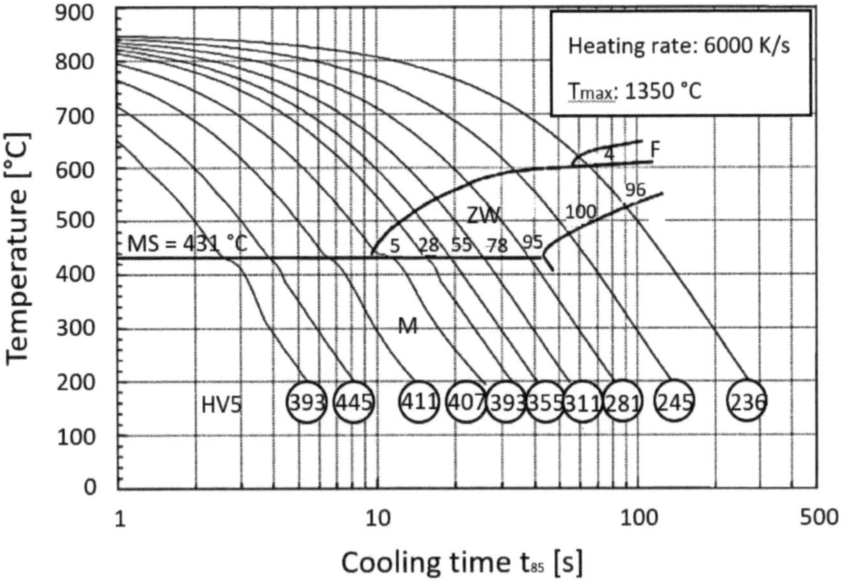

Figure 63: CCT diagram for steel S690 found in [12][8]

Table 32: Double parameter of the cooling material models for the simulation of FW690

CMM	2	3	4
T_{max} [°C]	817	882	1208
t_{85} [s]	4	4	4

Table 33: CMM 3 (T_{max} = 882 °C, t_{85} = 4 s) modelled for the simulation for the simulation of FW690

T	Microstructure					σ_{yi} [MPa]					σ_y	α^{se}
[°C]	PM	FE, PE	BA	MA	AY	PM	FE, EP	BA	MA	AY	[MPa]	[-]
850	31 %	0 %	0 %	0 %	69 %	62	61	40	73	48	53	15
719	31 %	0 %	0 %	0 %	69 %	150	94	101	246	79	101	15
431	31 %	0 %	0 %	0 %	69 %	594	237	332	750	134	277	15
200	31 %	0 %	0 %	63 %	6 %	676	312	386	853	200	762	12
20	31 %	0 %	0 %	63 %	6 %	734	472	394	977	245	861	12

[8] the diagram was partially redrawn by the author of the present dissertation in order to improve the figure quality

3.4.3.3 S960

Temperature-dependent thermal properties of S960 were considered identical to those of S690 presented in Table 28. Temperature-dependent parent material mechanical properties were extracted from the material characterisation of the investigated batch in the framework of the project *HFH-Simulation*. They are presented in Table 34. As in the case of S355 material characterisation took places in the range of 20 °C – 560 °C and the rest of the values were either inter- or extrapolated.

The chemical composition of the applied S960 batch and the results of the analytical calculation of the austenitization temperatures *A1*, *A3* and *A3'* are presented in Table 35 and Table 36 respectively. It was assumed that *A3'* = *A3*, as the calculated value of *A3'* was lower. 4 CMMs were modelled in order to simulate possible softening of the martensitic microstructure of S960 through austenitization and cooling. As no CCT diagram of S960 was found in literature, hardness measurements in the FZ and in the HAZ, which were carried out in the framework of the project *HFH-Simulation*, were used as input for modelling the CMMs. As is it shown in Figure 64 hardness of parent material, in the HAZ and in the FZ lies between 300 HV and 425 HV, with the highest values met in the HAZ and the lowest in the FZ. All values in this range indicate a very high proportion of martensite (analytical calculations with the present chemical composition result in a hardness of 342 HV for the martensitic microstructure). Therefore, the microstructural changes could have been neglected. Nevertheless, 4 CMMs were built in order to simulate these effects of hardening and softening in the HAZ and in the FZ respectively, based on the material properties of the individual phases (see Table 6). The values of the double parameter T_{max} and t_{85} characterizing each CMM are provided in Table 37. Exemplary, evolution of phase composition and the respective mechanical behaviour of CMM 4 are provided in Table 38. For the sake of shortness, the rest of the CMMs are provided in Appendix A.

Table 34: Temperature-dependent material parameters of S690 extracted in [146], used in mechanical analysis – Values are extrapolated for temperatures higher than 915 °C

T [°C]	σ_y [MPa]	E [GPa]	H [GPa]
20	1021	203	2.33
400	851	169	3.40
719	208	41	1.63
915	45	10	1
1500	7	10	1
3000	7	10	1

Table 35: Mean values of chemical composition for the S960 batch used in the project *HFH-Simulation*

C	Si	Mn	P	S	Al	V
0.22	1.25	0.008	0.001	0.22	0.052	0.043
Cr	Cu	Mo	Ti	Nb	Ni	
0.2	0.01	0.603	0.003	0.016	0.05	

Table 36: Calculated austenitization temperatures for the simulation of FW960

A1 [°C]	*A3* [°C]	*A3'* [°C]
719	1013	1013

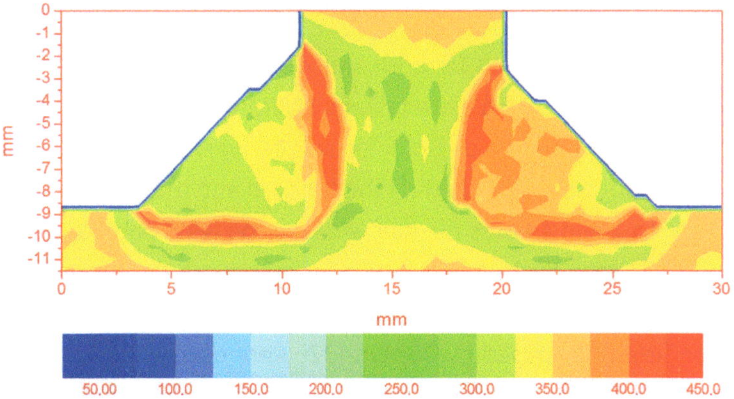

Figure 64: Vickers Hardness measurements (HV) carried out on FW960 by IWM in the framework of [146]

Table 37: Double parameter of the cooling material models for the simulation of FW960

CMM	2	3	4
T_{max} [°C]	866	964	1257
t_{85} [s]	4	4	4

Table 38: CMM 4 (T_{max} = 1257 °C, t_{85} = 4 s) modelled for the simulation of FW960

T	Microstructure					σ_{yi} [MPa]					σ_y	α^{se}
[°C]	PM	FE, PE	BA	MA	AY	PM	FE, EP	BA	MA	AY	[MPa]	[-]
1257	0 %	0 %	0 %	0 %	100 %	5	5	5	5	5	5	16
719	0 %	0 %	0 %	0 %	100 %	208	99	139	259	79	79	16
431	0 %	0 %	0 %	0 %	100 %	826	251	457	787	134	134	16
200	0 %	0 %	0 %	92 %	8%	940	331	531	895	200	839	12
20	0 %	0 %	0 %	92 %	8%	1021	500	542	1025	245	963	12

3.4.4 Analyses and Results

3.4.4.1 Thermal Analysis

Thermal analysis was carried out based on the above-presented theoretical background. Two different analyses were carried out for FW355 and for FW690/FW960 respectively. A single analysis was carried out for the two high strength steels, as they exhibit identical thermal behaviour and were welded with the same WPS.

The heat source was calibrated for both analyses based on macrosections of the investigated fillet welds, which were taken by IWM during *HFH-Simulation* [146]. Initially, the width and depth of Goldak's source were adapted directly to the measured weld pool size. The length of the forward and rear quadrant were once again set equal to 1 and 4 times the half width of the weld pool respectively. The macrosections along with the measured dimensions are displayed in Figure 65. The applied welding parameters are presented in Table 39. Values for the weld metal arc coefficient proposed by Dupont [32] were used once again. Still, initial thermal results have shown that this setup leads to a smaller weld pool size in the simulation, i.e. temperature of some elements inside and near the boundaries of the FZ did not reach T_{melt}. Consequently, the parameters of the heat source had to be recalibrated.

Through a *trial and error* procedure the parameters of the Goldak's source were reset so that the FZ in the simulation would match the real one and the agreement of the calculated thermal profiles with the respective thermocouples' measurements, which were carried out in the framework of *HFH-Simulation* [146], would be sufficient. Finally, a common width of the weld source of 6 mm, an equal depth and a retained ration of 4:1:1 for the forward and rear length to the half width were selected. The final parameters of the heat source are presented as well in Table 39. The calculated FZ and the temperature distribution in the fusion zone at the middle of the heat source were compared with the real macrosection after recalibration. This comparison is illustrated in Figure 66. On the left side of the figure, the macrosections are presented for both materials with the boundaries of the FZ being highlighted with a black line. At the middle, qualitative contours of the temperature distribution are presented. Exemplary, the quantitative contour for the component FW355 (mirrored) is presented in Figure 67.

Table 39: Initial and final parameters of Goldak's source applied for the thermal analysis

Goldak's Source	Initial Setup		After Calibration	
	S355	S690/S960	S355	S690/S960
a [mm]	4.85	4.94	6	6
c [mm]	7.87	4.94	6	6
b_f [mm]	4.85	4.94	6	6
b_r [mm]	19.40	19.80	24	24
η [-]	0.85	0.85	0.85	0.85

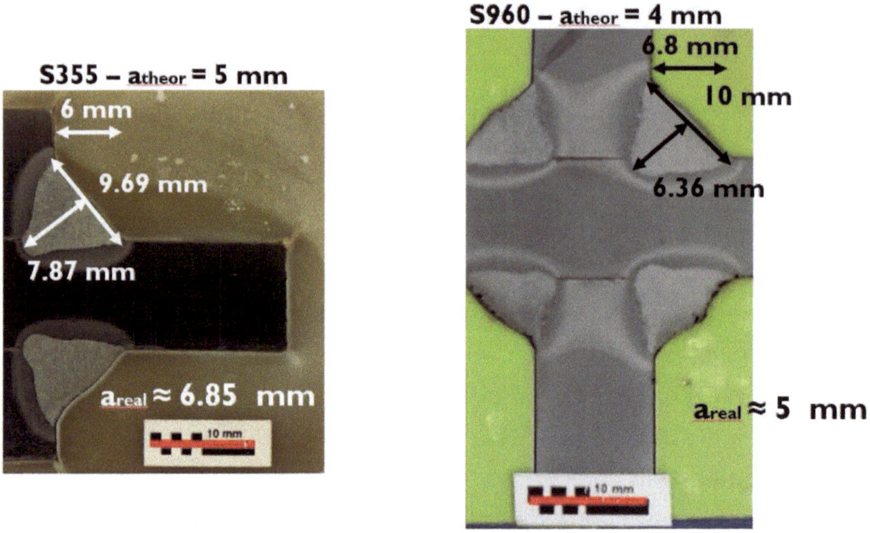

Figure 65: Macrosections of the investigated fillet welds taken by IWM [146] – Measurement of the weld pool size

On the right hand side of Figure 66 the FZ and the HAZ are presented, as they were calculated from the applied algorithm, along with the boundaries of the real FZ. Comparing the real and the simulated ones, a very good agreement to the shape and the width for both the FZ and the HAZ is met for S355. A small underestimation of the depth of the FZ is observed though. In the case of S690/S960 on the other hand the width of the source at half depth is overestimated.

Regarding the quantitative contour of temperature in the heat source transverse to the centre in longitudinal direction at its centre in Figure 67, the temperatures at the predicted boundaries of the weld pool and in the HAZ confirm the expected values of T_{max}. The temperature at the centre of the weld pool however, is approaching the boiling temperature of iron (≈ 2870 °C). Previous investigations on austenitic steels have shown that the maximum temperatures inside the weld pool can be higher than 3000 °C depending on the welding method and material [97], [179]. Albeit, in ferritic and martensitic steels maximum temperature is expected to be lower, due to the lower requested thermal input.

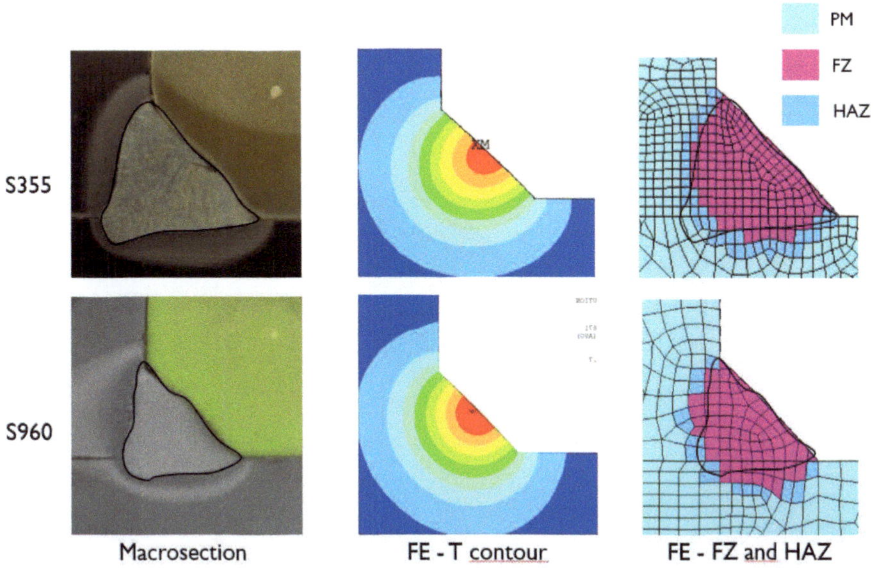

Figure 66: Macrosections of the investigated fillet welds taken by IWM [146] – Measurement of the weld pool size

The above-mentioned discrepancies are assumed to be negligible regarding the calculation of WRS. The negligible influence of small deviation in the size of the heat source was proven in the previous investigations of single-pass butt welds. As long as the heat flux inside the component, especially from the boundaries of the FZ up to the PM close to HAZ, is calculated with sufficient preciseness, the influence of erroneous maximum temperature at the centre of the weld pool is considered to be negligible. In any case, during the mechanical analysis all elements are deactivated when they exceed the melting temperature, so this could only influence the energy input in the component. Still, very good agreement is met though between the calculated temperature profiles and the respective measurements, which were carried out by means of thermocouples. The comparison for S355 and S690/S960 are presented in Figure 68 and Figure 69 respectively.

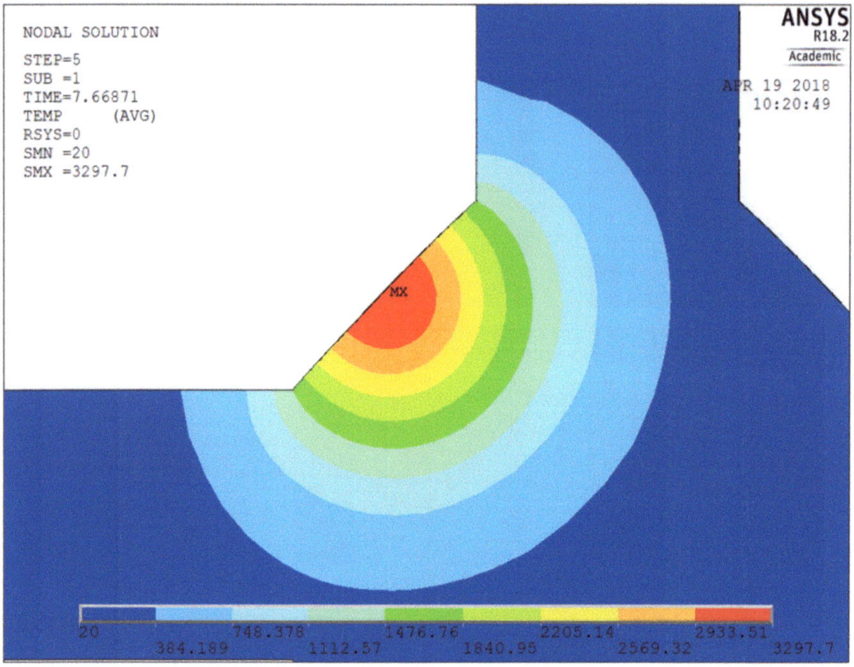

Figure 67: Quantitative temperature contours in the weld zone – Temperatures in °C

In both cases, T_{max} is calculated with very good preciseness. Heating and cooling rates are slightly overestimated. In any case, the heating rate is insignificant for the WRS with the current modelling approach. As for the cooling rate, observing the CCT diagrams of Figure 62 and Figure 63 a small overestimation of 2 s of the cooling time leads to negligible difference in the formatted microstructure during cooling. The neglection of fluid flow and the respective convective heat transfer inside the weld pool is considered to be the reason for these negligible discrepancies in the thermal analysis.

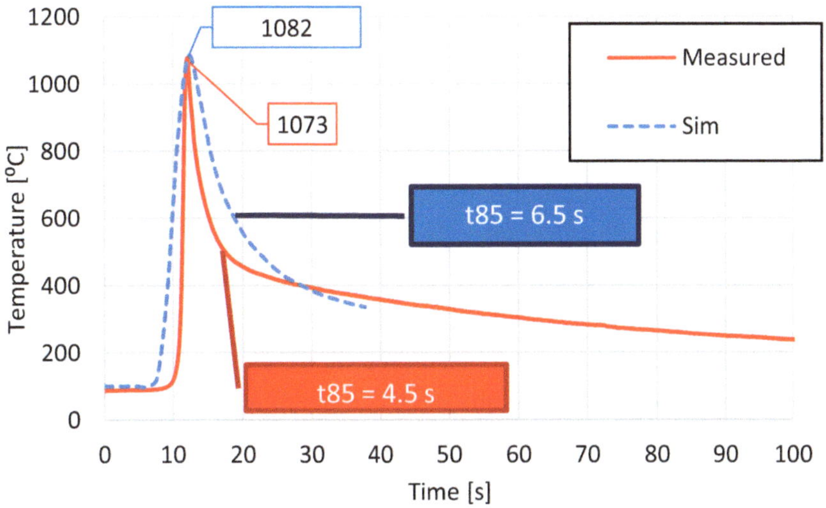

Figure 68: Comparison of the thermal analysis and thermocouples measurements of FW355

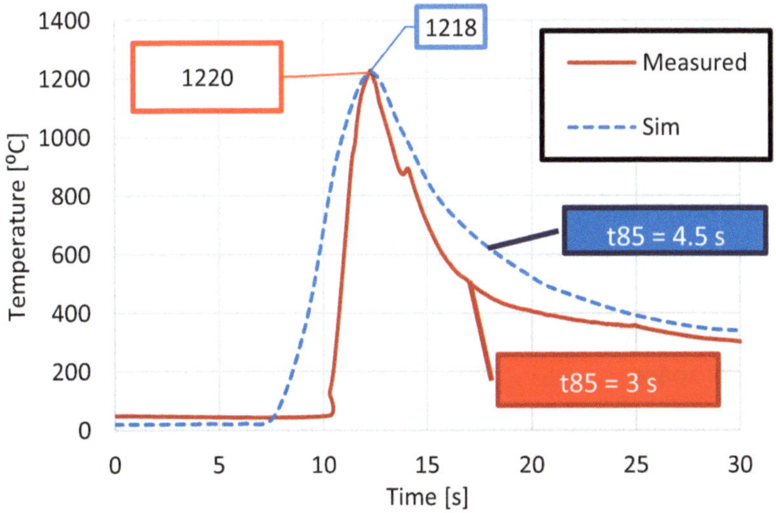

Figure 69: Comparison of the thermal analysis and thermocouples measurements of FW960

3.4.4.2 Structural Analysis

Two different levels of simulation were applied out for the structural part of the present analysis. One using simple boundary conditions was solved in the framework of *HFH-Simulation* and one more sophisticated based on the above-presented concept of *stiff springs* in the framework of the present dissertation. It was initially intended to upgrade the modelling approach further by considering the Bauschinger effect for the material away from the weld, which is not influenced severely by the thermal input, additionally to the application of *stiff springs*. Nonetheless, due to convergence difficulties, isotropic hardening behaviour was applied for the whole components. Furthermore and for the same reason, large deformations (non-infinitesimal strains) were not considered during the *present* solution with *stiff springs*. The influence of T_{ref}, the setup of the clamping mechanisms and the specimen cutting on WRS were investigated as well during this series on analyses. The WRS measurements from HFH-Simulation were applied for validation. Although, measurements were carried out on the 1st pass and the respective numerical results were compared at first, an overall comparison of the results from all passes was considered valid; the discrepancies of the measured WRS and signs of local effects led to this decision. Hence, the WRS were mirrored on both sides of the WRS diagrams (left and right to the stiffener) so that a direct comparison of numerical results to the measurements could be enabled for all passes.

Reference Temperature of Resolidified Material

The above-stated problem of T_{ref} had to be resolved before proceeding to further analyses. For this reason, the WRS, computed with and without reversal of T_{ref}, were compared with the measured WRS for all three investigated components, FW355, FW690 and FW960. As this comparison provided similar results in all cases, only the results regarding S355 are presented in the current dissertation for the sake of space.

The simulated longitudinal and transverse WRS for the component FW355 along with the respective mean of the measurements are presented in Figure 70 and Figure 71 respectively. The profiles of the first three passes calculated without reversal of T_{ref} validate the theoretical ones, exhibiting tensile residual stresses in the region of the weld seam and the weld toe and passing in the compressive region away from them. Results are different for the WRS of the 4th pass, especially in the case of transverse stresses, due to the thermal influence of prior passes, an effect which will be discussed below. It has to be mentioned that the theoretical profiles predict a counterbalance between the tensile and the compressive areas but this should stand for membrane stress profiles. As the present profiles are taken on the surface of a three dimensional component with non-negligible thickness, such equilibrium interpretations should be made on the integrated sums of forces i.e. on the whole thickness of the component.

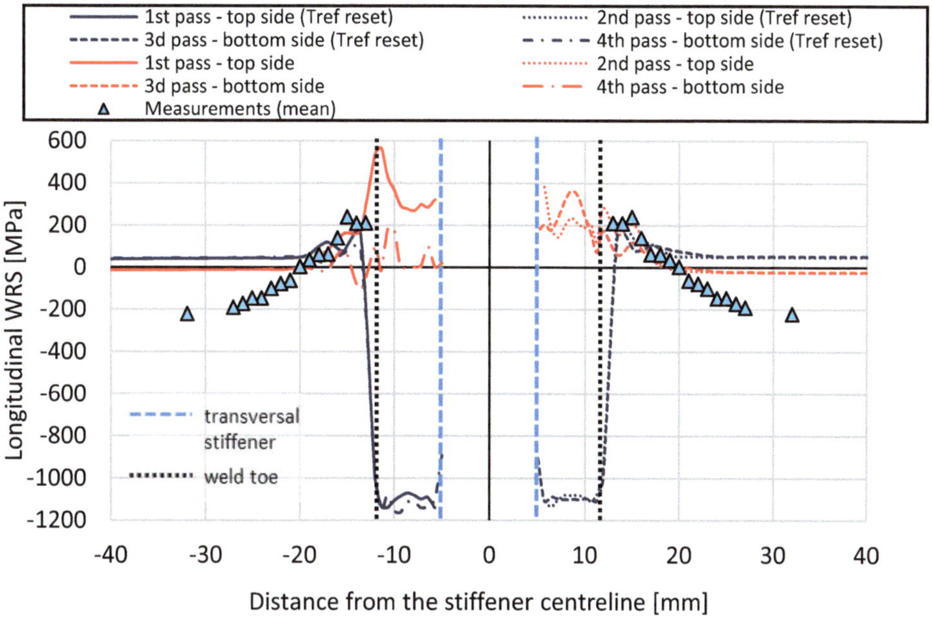

Figure 70: Longitudinal WRS at the centre of component FW355 – Influence of T_{ref}

On the other hand and as in previous investigations (see Figure 51 and Figure 52), the model considering reversal of T_{ref} produces diagrams with compressive stresses in the weld area. Although it is against engineering common knowledge, which dictates the presence of tensile stresses in this area, this is possible in some cases due to microstructural transformations [40]. Nonetheless, the agreement that the model with non-reversed T_{ref} exhibits with measurements, near the weld toe, is very good and especially in the case of longitudinal stress, whereby the measurements are considered more robust. On the other hand, the calculated profiles of the model with reversed T_{ref} did not seem to follow the real RS profiles. Hence, the model excluding reversal of T_{ref} was selected for all subsequent simulations. The deviation between the simulated and measured WRS in the region away from the weld is discussed below.

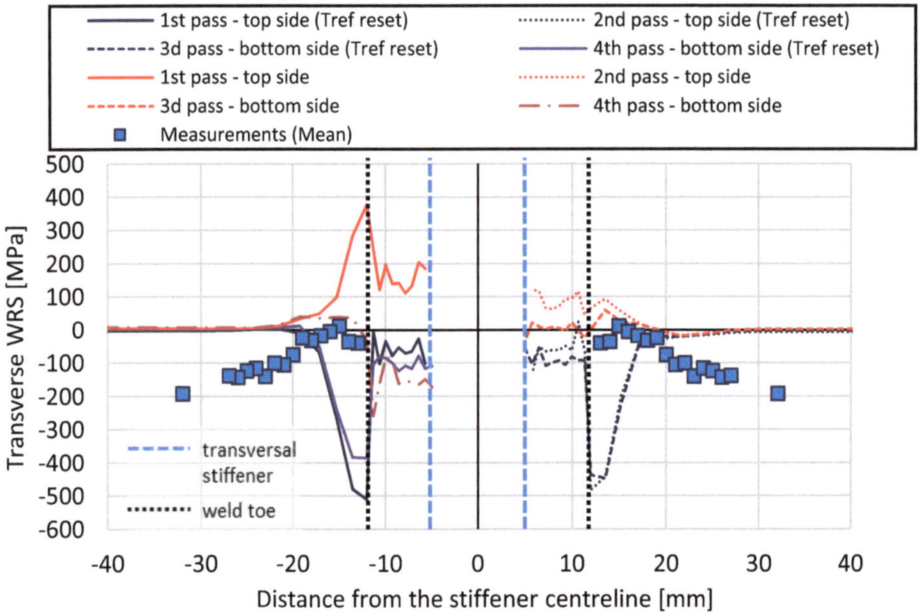

Figure 71: Transverse WRS at the centre of component FW355 – Influence of T_{ref}

Overall Validation

The longitudinal WRS for all three investigated components, calculated with the modelling approach of *HFH-Simulation* (tagged *"HFH Sim"* - fixing of the clamped nodes) and the current one (tagged as *"present"* – using *stiff springs* and improved material modelling) are presented in Figure 72, Figure 73 and Figure 74 along with the respective measurements. Similar results are met for all materials and for both modelling approaches. The highest and the lowest WRS are met on the 1st and the 4th passes respectively. In the case of FW355, tensile stresses around 630 MPa clearly higher than the nominal yield strength are met on the weld toe of the 1st pass. They match though the order of magnitude of the higher yield strength in the HAZ, which is increased due to microstructural transformations.

The large deviation to the tensile WRS of the 1st to the rest of the passes can be attributed to the effects caused by the thermal influence of prior to consequent weld passes. No specifications were given regarding the intermediate temperature in the WPS, so probably all passes were welded consecutively. Therefore, a cooling time between passes of 120 s was assumed[9]. At this time, temperature in the weld region is approximately 200 °C (see Figure 69 for example). Thus, heat input from the prior weld passes acts as preheating for the former ones, reducing in this way the introduced WRS. As no preheating is applied prior to welding, the first pass is cold-welded and exhibits significantly higher WRS than the rest of the passes as expected.

Furthermore, as cooling time was similar for all passes, T_{max} was predominant for the rest of them. Heat input from passes 1 and 2 acted as preheating for the passes 4 and 3 respectively and higher T_{max} was reached. This led to assignment of different microstructural models to the first two and later passes. CMM 4 and 2 were assigned to the HAZ of the 1st and 2nd passes exhibiting respectively yield stresses of 947 MPa and 586 MPa. CMM 3 with yield strength of 766 MPa was assigned to the other two instead of CMM 2. Therewith, a broader zone of high strength is created allowing for no sudden peak of strength on the weld toe that could be secondary accounted for the highest peak of WRS on the first pass. Exemplary, the CCM assignment for the 1st and 4th pass of S355 is presented in Figure 75. This effect was limited for the case of the two high strength steels due to the homogeneity when yield strength is considered. Small changes in yield stress due to microstructural transformation took place.

The influence of modelling approach for the boundary conditions on the tensile WRS, as expected, is limited for all three components. Application of *stiff springs* produces slightly higher WRS on the weld toe and, due to equilibrium, slightly lower away from it.

---

[9] no documentation available, 120 s were assumed rational for practical reasons i.e. for the welder to change position

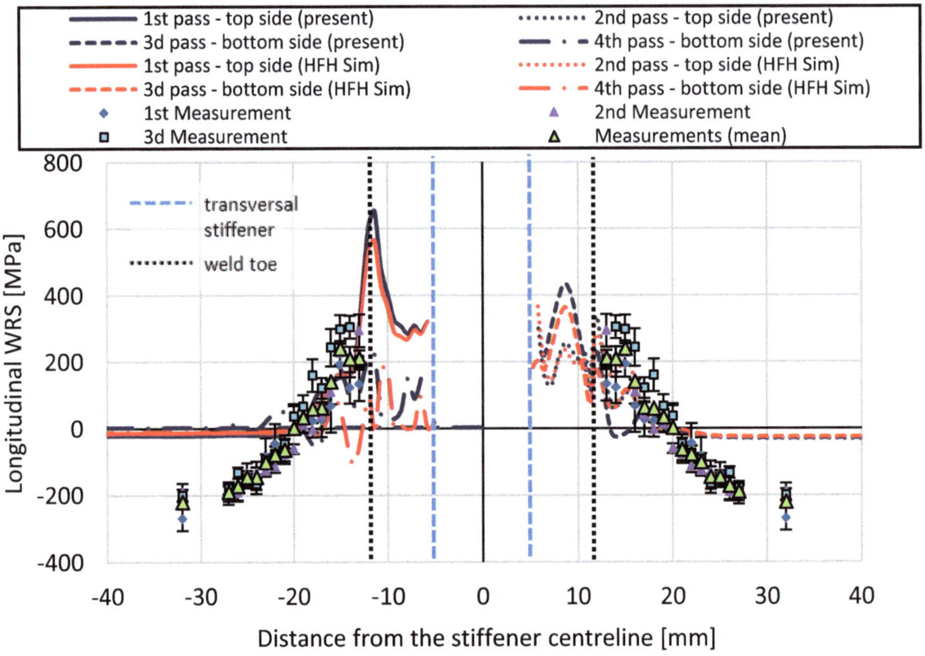

Figure 72: Longitudinal WRS at the centre of component FW355 – measured WRS from the 1st pass are mirrored on the right side

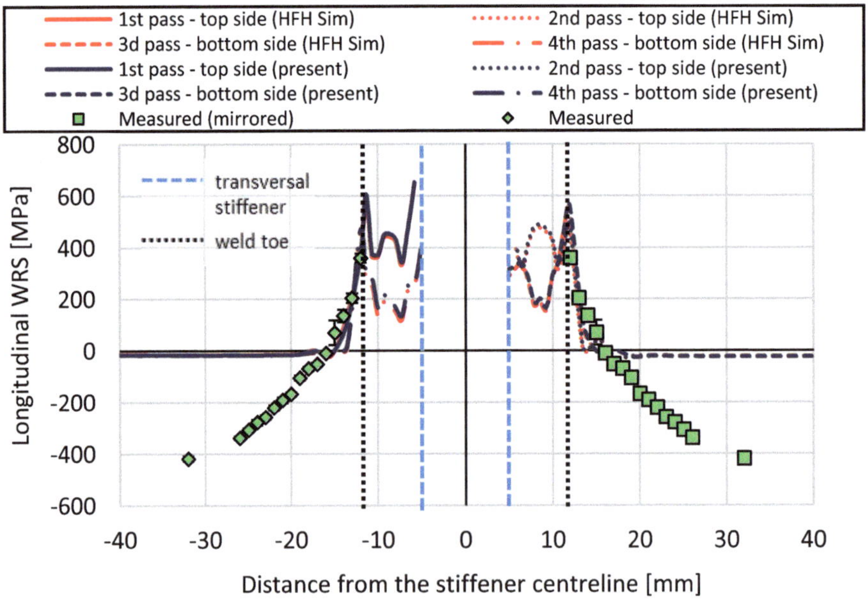

Figure 73: Longitudinal WRS at the centre of component FW690

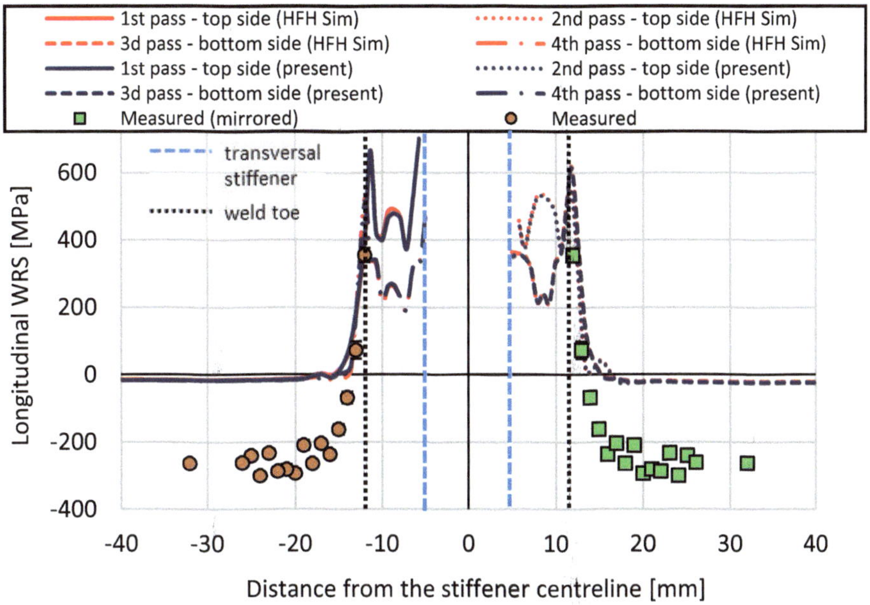

Figure 74: Longitudinal WRS at the centre of component FW960

The overall agreement of the simulated and measured longitudinal WRS on the 1st pass is good in all cases for a region of up to 5mm away from the weld toe. In this region, the other simulated profiles match the measured ones as well. Away from this region of the 5 mm the compressive stresses introduced by the manufacturing processes (rolling or shot blasting) cause a significant deviation between simulation and measurements.

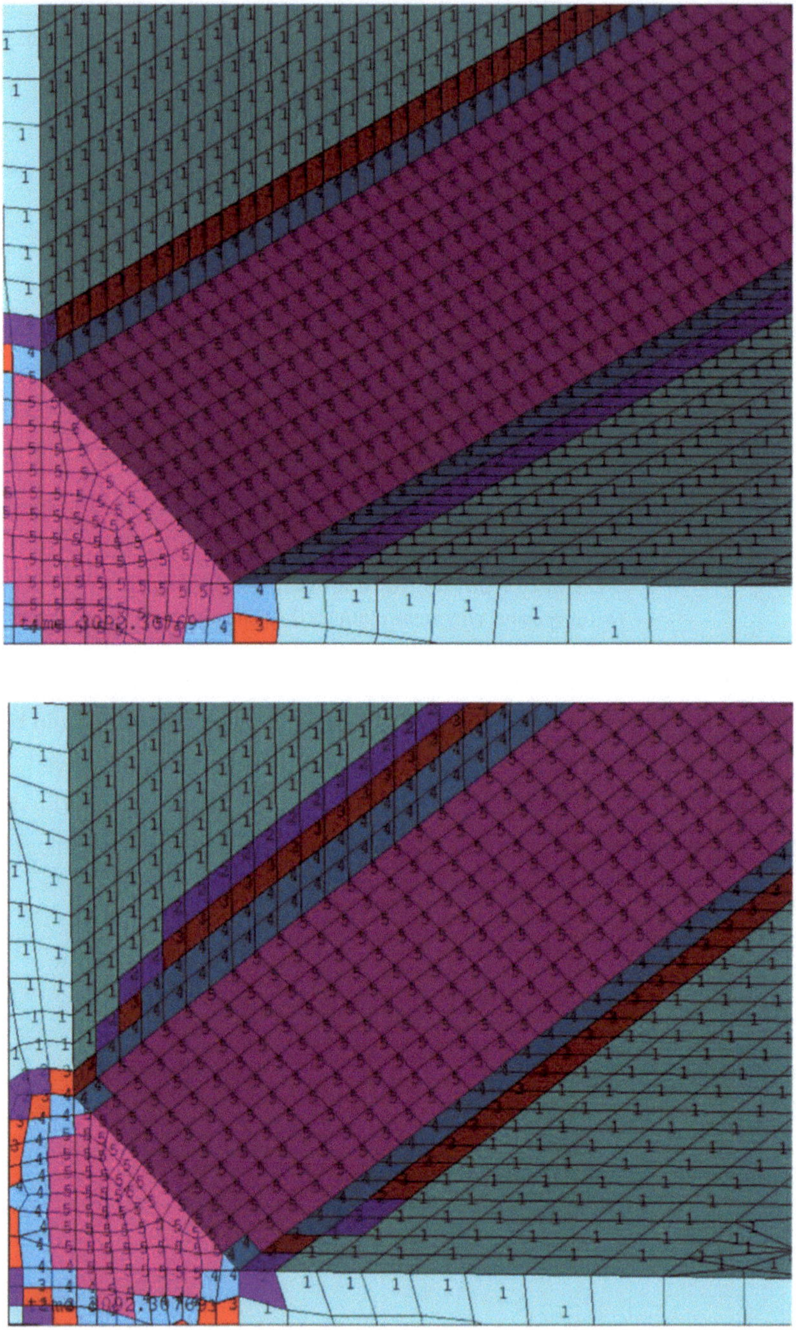

(a)

(b)

Figure 75: CMM assignment for S355 (CMM 5 identical with CMM 4 but with T_{ref} = T_{melt}, Material 1 is PM): a) 1st pass; b) 4th pass

The calculated transverse WRS for all three investigated components, based on the simple modelling of boundary conditions applied for *HFH-Simulation* are presented in Figure 76, Figure 77 and Figure 78. Similar results are met once again in all cases. The highest WRS are met on the 1st pass and the lowest on the 4th, with the extreme case found in component FW355, where compressive stresses are calculated in the latter. The same above-described reasons for this deviation of longitudinal stresses should stand in this case once again.

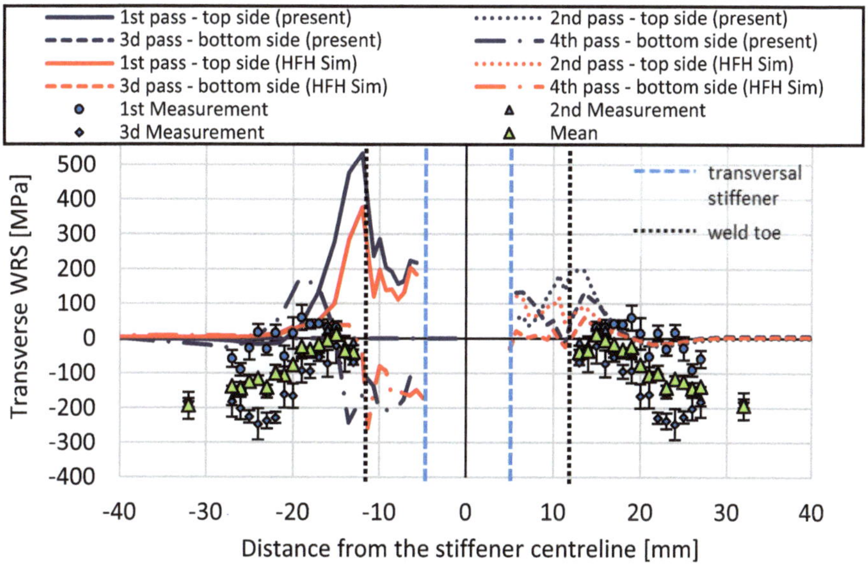

Figure 76: Transverse WRS at the centre of component FW355

For all three materials, the magnitude of the calculated transverse WRS is smaller than the longitudinal. For the component FW355 the profile of the 1st pass is lying inside the scatter band of the measurements near the weld toe. Deviation in the region away from it is once again attributed to the influence of the RS introduced by the manufacturing processes. On the contrary, the calculated transverse WRS for the other two components, FW690 and FW960, deviate completely from the measured WRS. The arbitrary assumption of *FWBC1* boundary conditions for the components FW690 and FW960 could as well be causing this deviation[10]. The agreement of the present simulation in the case of longitudinal stress, at least at the area close to the weld, along with the great uncertainty of the measured WRS,

[10] during the project *HFH-Simulation* it was not documented on which side of the specimens (1st and 4th pass or 2nd and 3d pass side) the clampers were applied – *FWBC1* was assumed for the rest of the simulations

as it is exposed from the multiple measurements on S355, could as well arise questions regarding the validity of the measured transverse WRS.

In the case of S355, the influence of modelling approach on the transverse WRS, as expected, is more significant than for the longitudinal. Application of *stiff springs* produces WRS of significantly larger magnitude near the weld toe, as in the case of the previously investigated butt welds. For the other two materials negligible difference is observed. The unavoidable neglection of geometrical non-linearities (large strains were neglected), which was previously proven predominant for the transverse WRS due to convergence problems, could be accounted for this. The broad scatter band of the WRS measurements for S355 and negligible difference between the *HFH-Simulation* and the *present* approach though, do not allow for exclusion of one of the two methods.

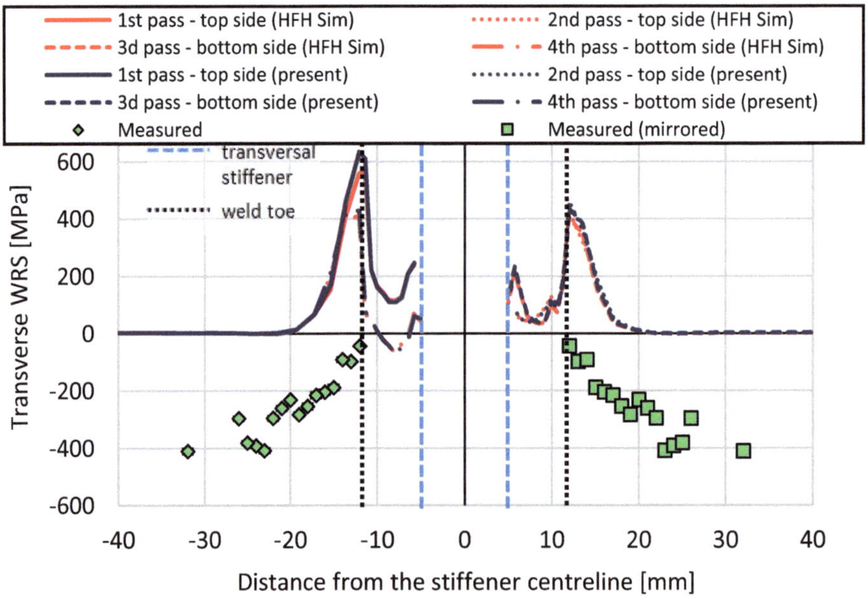

Figure 77: Transverse WRS at the centre of component FW690

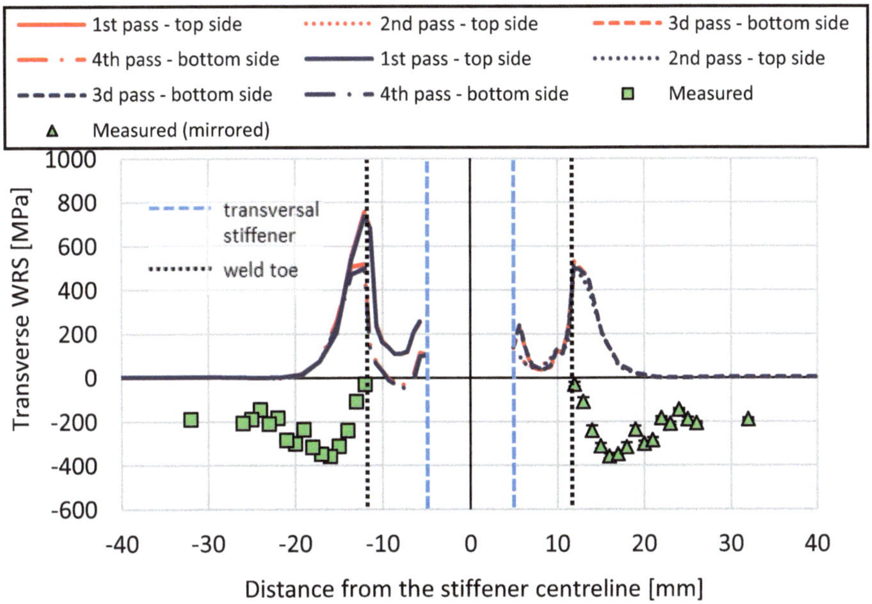

Figure 78: Transverse WRS at the centre of component FW960

Influence of Symmetric Automated Welding of High Strength Steels

The influence of welding sequence in the case of high strength steels was investigated by simulating the symmetric welding sequence, which was applied for the real fillet welds of the project *HFH-Simulation*. The results were compared with the assumed simulated sequential welding (tagged above as *present*), identical with the real welding sequence of the specimens of S355. Identical welding parameters were applied in both cases. The longitudinal and transverse RS for the component of S960 simulated with the two sequences, sequential and symmetrical, are compared with each other and with the measured RS in Figure 88 and Figure 97 respectively. The two investigated sequences produce overall significantly different WRS profiles. The symmetric setup produces significantly lower RS in the two upper passes, which were welded first. In the case of the transverse RS, similar conclusions are drawn from the comparison between the two methods and a significant deviation from the measured RS due to the previously stated reasons is once again evident. Actually, the difference between the WRS peaks calculated with the two sequences becomes even more significant. As sequential welding produces the highest peak of both longitudinal and transversal WRS, it is the less favourable and is adopted for all subsequent simulations. Similar results were met for the components of S690 as well and they are omitted for the sake of space.

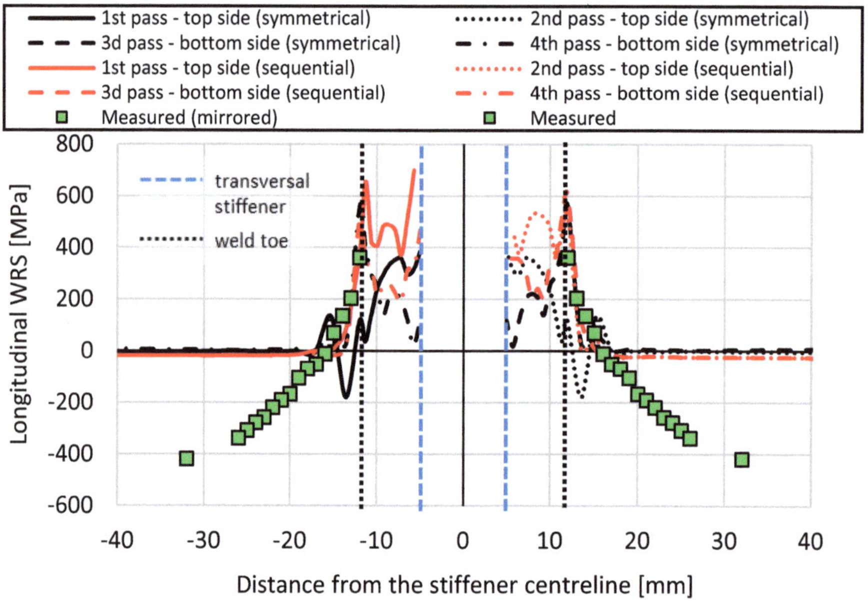

Figure 79: Longitudinal WRS at the centre of component FW960 – Influence of symmetric welding

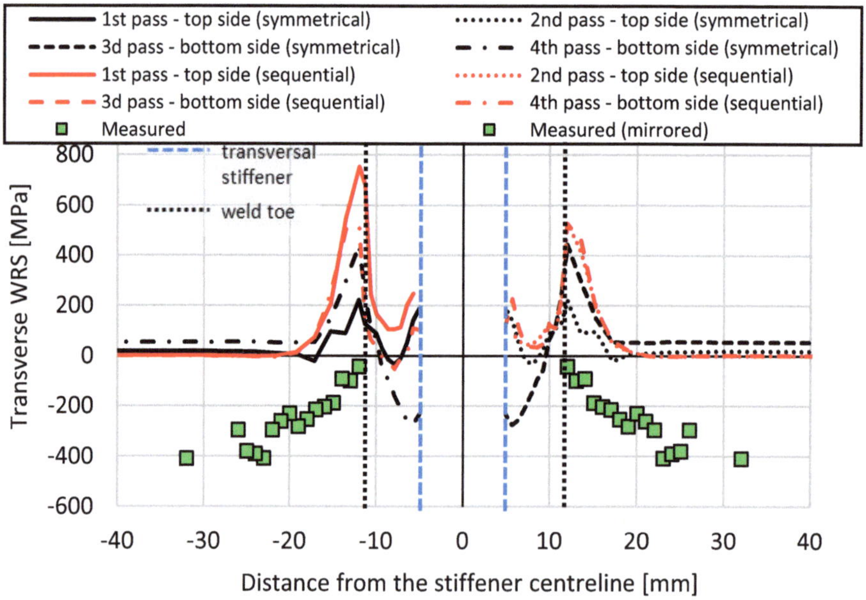

Figure 80: Transverse WRS at the centre of component FW960 – Influence of symmetric welding

Influence of Clamping Setup in the Case of Fillet Welds

The influence of different boundary conditions in the case of fillet welds was investigated for all components based on the modelling approach from *HFH-Simulation*. The present setup *FWBC1* was compared with two alternatives. In *FWBC2* the clampers were assumed to be on the opposite side of the plate (side of the 2nd and 3rd passes). In the case *FWBC3* clampers were applied during the first two passes on the left side (side of 1st and 3rd passes, as it is illustrated in Figure 56) and during the latter two on the opposite side of 2nd and 4th passes. The same investigations were repeated with *present* modelling. As analogous results were calculated in all cases and with both modelling approaches, only the results regarding the component FW355 and calculated with the method *HFH-Simulation* are presented here.

The longitudinal and transverse WRS calculated with each boundary condition setup are presented in Figure 81 and Figure 82 respectively. The influence of the clamping setup on the longitudinal stresses is negligible as expected. *FWBC1* and *FWBC3* produce almost identical profiles, while they exhibit slight deviations with the setup *FWBC2*. Still, these deviations are up to 50 MPa and are not found on the locations of peak stresses so they can safely be neglected. On the other hand and in the case of transverse stresses, non-negligible deviations are met. Once again, *FWBC1* and *FWBC3* setups produce almost identical profiles overlapping with each other, but the deviation with the case *FWBC2* cannot be neglected anymore. Simulation *FWBC2* produces results with stress peaks of significantly larger magnitude, of even up to 100 MPa, in all welding passes. This setup is therefore considered inefficient.

The major similarity between *FWBC1* and *FWBC3* that contributes to the introduction of identical stress profiles is the restraining of the left side of the plate (side of the 1st and 4th pass) during welding of the 1st pass. As the highest stresses are met on this pass, the influence of the external restraints on it seems to affect all consecutive welding passes due to equilibrium reasons. It can be safely assumed that restraining of the opposite side from the one currently under welding leads to less stiff restraining and in extension to lower WRS. Therewith, textbook knowledge regarding external restraints and their influence on transverse stresses is confirmed. Especially for the cold welded 1st pass, where the highest tensile WRS are met, this effect is predominant

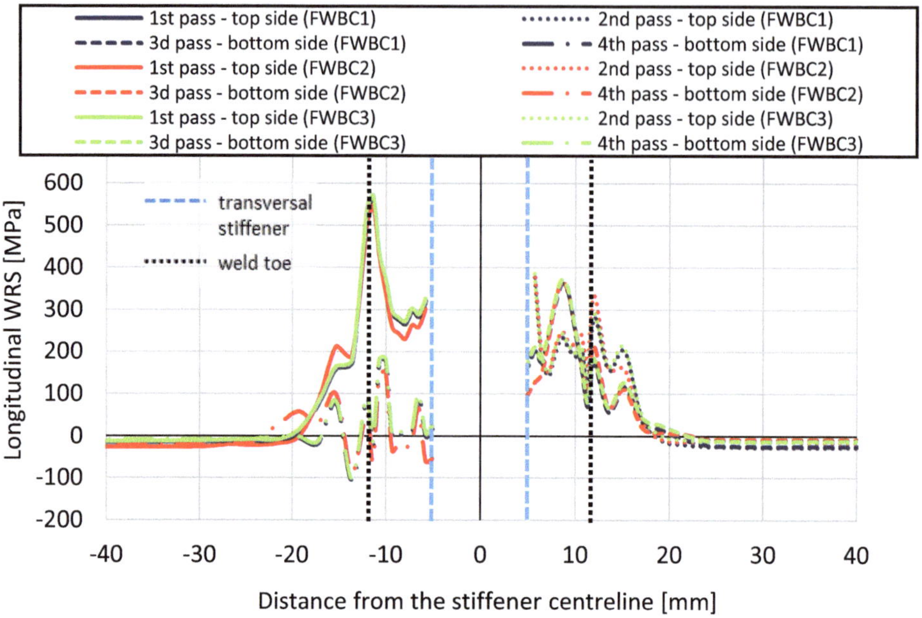

Figure 81: Longitudinal WRS at the centre of component FW960 – Influence of clamping setup

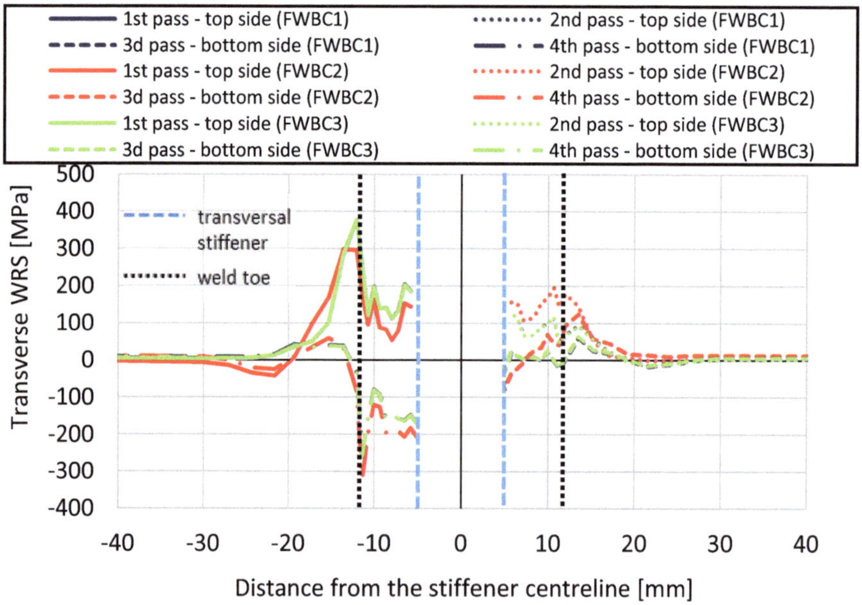

Figure 82: Transverse WRS at the centre of component FW960 – Influence of clamping setup

Influence of Shot Blasting prior to Welding on the RS Profile

In order to account for the effect of preceding shot blasting on the WRS profile, a pseudothermal approach was proposed in [53], a study that was carried out in the framework of the present doctoral dissertation. As a full simulation of the shot blasting mechanical event would not be numerically feasible and keeping in mind that this would diverge from the main objective of the current work, applying a thermal load to introduce the desired compressive field was considered instead. The main idea was to reverse the effect of temperature during welding. Instead of applying concentrated high temperature on the middle of the investigated plate, a negative temperature would by introduced, using Equation (18). It was initially expected to be a quick and numerically efficient approach. Moreover, as it was described above, significant deviation was met for the compressive WRS measurements attributed to shot blasting. Therefore, no precise RS profile existed to be simulated.

The first step of this side-path study was to simulate on a small component of a 20MnCr5 steel the RS profile for shot blasting from Figure 15, as it was measured by Shaw et al. [149]. A rational, initial assumption was to apply a through depth temperature distribution, which would qualitatively follow the profile of the measured RS. In this case, a significant differentiation of the RS in the first 100 μm below the surface was documented. Consequently, a significantly fine mesh of 0.0025 mm was applied on the top layer and it was constantly coarsening in through-depth direction. The magnitude of this temperature profile was changed so that the influence on the RS could be documented. The results were satisfying in the sense that a very quick simulation method, directly applicable to the existing welding model enabled the creation of a compressive stress field on the applied area of the component. Nevertheless, due to the nonlinear nature of the material, the RS profile could not be simulated with preciseness. More details about these preliminary analyses and the applied method are omitted for the sake of space and can be found in [53]

Following, implementation of the pseudothermal loads on the fillet welds of S355 preceding to the mechanical solution of the weld model was carried out. In this case, as the mesh should be common with the weld simulation, no such sophisticated application of thermal loads in through-depth direction would be possible and therefore the minus temperature of -500 °C was applied only on the top and bottom surfaces. The influence on the calculated WRS is presented for the first time in Figure 83 of the present study. A shift of the RS profile downwards was indeed documented. Nonetheless, this shift was met near the weld toe and not in the adjacent area, where the compressive stresses were measured. Different application of boundary conditions or thermal loading could lead to an improvement of the results. Further analysis though was excluded, as it would surpass the boundaries of the present study. Moreover, the agreement of the above-presented weld simulations is very good in all cases at the regions of interest, i.e. near the weld toe and in the HAZ area, where the effect of shot blasting is eliminated due to the thermal treatment of the material during welding. Therefore, further investigation was considered obsolete for the purpose of the present study.

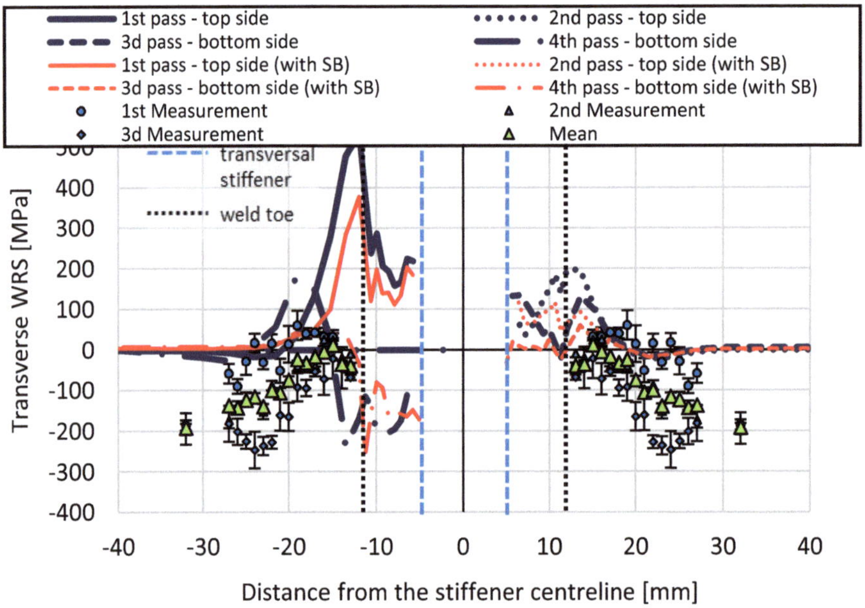

Figure 83: Influence of introducing shot blasting pseudothermal modelling in the present weld simulation

Cut Specimens

Stress relaxation due to specimen cutting was simulated by deactivating all the elements outside the specimen contour and erasing all external restraints and solving one last, additional step at the end of the structural quasi-static analysis. Due to the non-orthogonal shape of elements in the mesh transition area a width of 70 mm was selected, while specimens of 50 mm were cut in *HFH-Simulation*.

Such a scaling effect was considered negligible. Similar pattern of stress relaxation was met in all cases. For the sake of space, only the results for S960 and the present modelling approach are currently presented. The contour of transverse stresses at the prior and the last step of the solution, showing the state of stresses on whole plate FW960 and on a cut specimen respectively, are presented in Figure 84 and Figure 85. A homogenous compressive stress field away from the weld toe is met in both cases, while tensile stress are met in both cases in the weld area. A significant stress relaxation is observed when the two contours are compared both in the tensile and in the compressive region without any obvious qualitative changes.

A better overview is possible when the profiles of longitudinal and transverse WRS transverse to the weld line, which are presented in Figure 86 and Figure 87 respectively, are considered. For both profiles, transverse and longitudinal, no qualitative change is caused but a significant shift downwards is obvious. As expected, the relaxation for the longitudinal stresses is significant, while for the transverse it can be considered negligible. The stress profiles for both transverse and longitudinal WRS along the weld toe of the specimen's 4th pass were as well considered and they are presented in Figure 88. In this case, a change to the shape of the WRS profile is met as well. Prior to cutting, homogenous fields are observed as a specimen from the middle of the plate is considered. The stress relaxation after cutting is however more significant on the edges of the cut specimen, validating initial expectations. For the transverse WRS, a small increase is met at the centre of the specimen due to equilibrium reasons; the stress relaxation at the edges of the specimen is small and therefore the stress at the middle is slightly increased. Overall, the numerical results validate at least qualitatively initial expectations, still without producing compressive longitudinal stresses as in Figure 18. Further investigation, comparing measured and numerically calculated WRS from the same specimen are proposed.

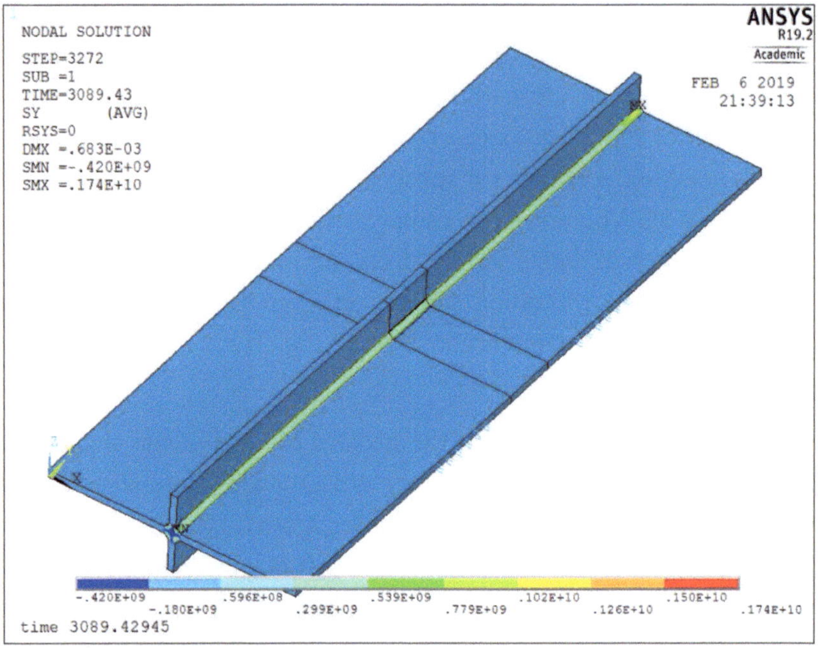

Figure 84: Transverse WRS of the whole plate FW355, stresses are given in Pa – Contour of the single specimen is marked with black line

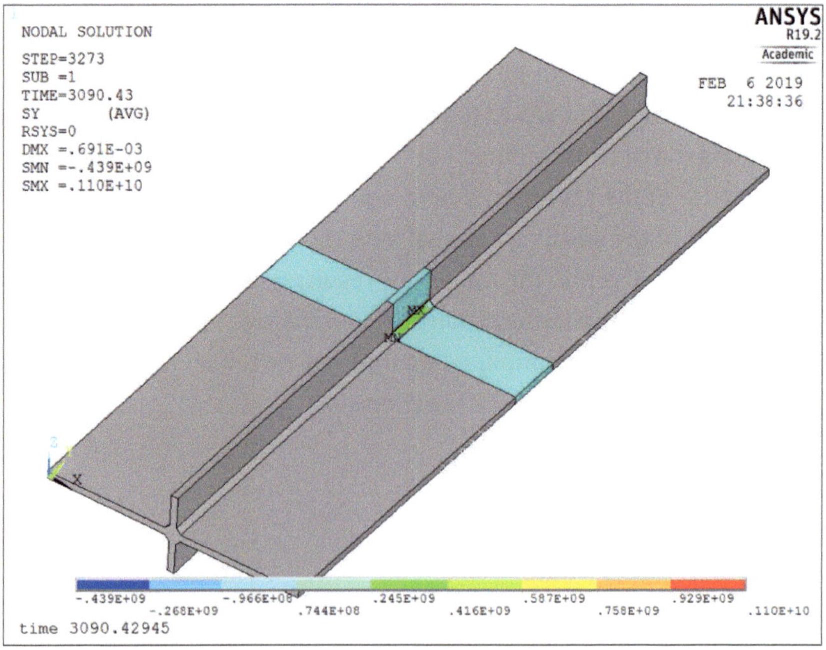

Figure 85: Transverse WRS of the cut specimen from FW355, stresses are given in Pa – Area of deactivated elements are presented in grey

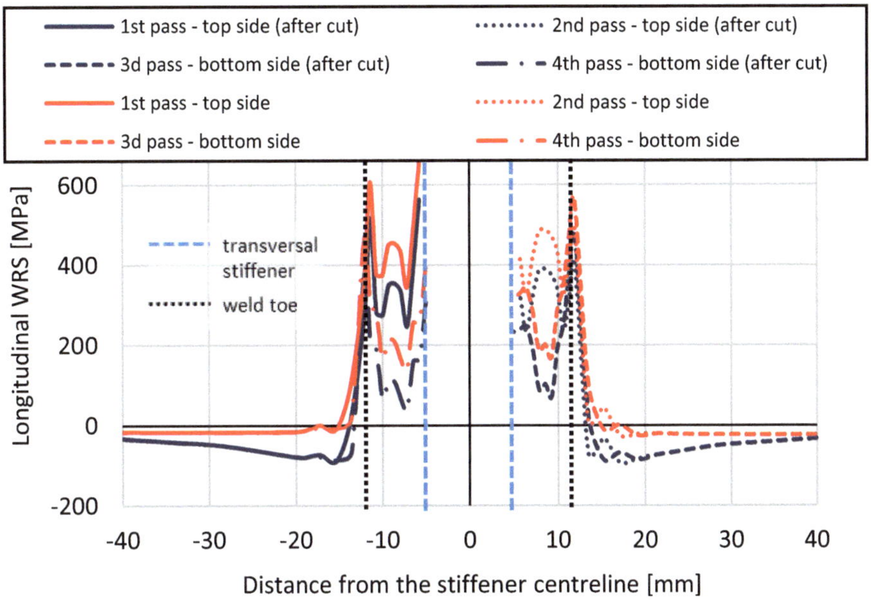

Figure 86: Stress relaxation of longitudinal WRS due to cut of specimen from component FW960 transverse to the weld line at the centre of the component

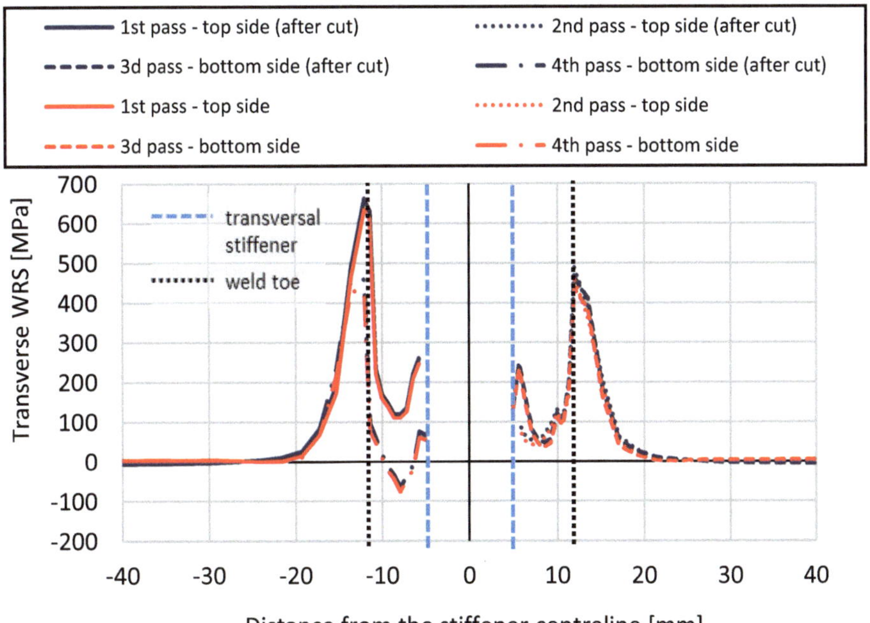

Figure 87: Stress relaxation of transverse WRS due to cut of specimen from component FW960 transverse to the weld line at the centre of the component

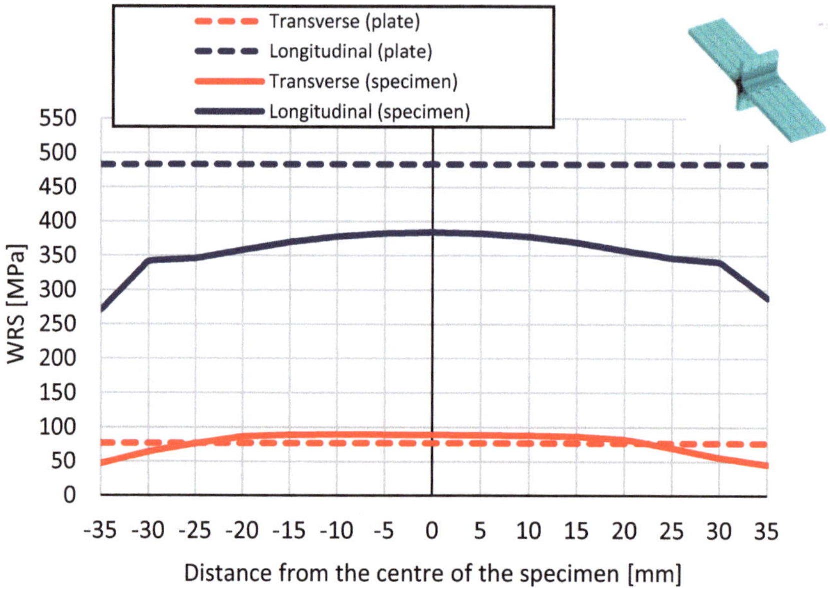

Figure 88: Stress relaxation of WRS due to cut of specimen from component FW960 along the weld toe of the 4th weld pass

3.4.5 Conclusions

A series of analysis was carried out regarding the WRS fillet welds with double-sided transversal stiffeners. This geometry was selected, as it was investigated in the project *HFH-Simulation* and WRS measurements in as-welded and HFMI-treated state were available. Hence, validation of both welding and HFMI simulation could be carried out. Results of weld simulation would be the input for the subsequent HFMI Simulation. The previously validated engineering approach was applied with modifications wherever mentioned. When preciseness is considered, two levels of simulation were carried out. The former in the framework of *HFH-Simulation* [146], regarding simple modelling of boundary approach, and the latter one was presented for the first time in this dissertation. The following conclusions were drawn:

- The proposed engineering approach provides good results in the case of fillet welds as well and is then universal with regard to geometry.
- WRS measurements can locally exhibit significant deviations at nearby locations of a component due to manufacturing processes. This effect should

always be considered, when these measurements are applied for the validation of numerical results, which may not be able to depict these deviations.

- The rolling and/or the shot blasting process, which is usually applied for cleaning of metallic surfaces prior to welding, introduced significant compressive residual stresses. These influences cannot be ignored when WRS are simulated overall. Nevertheless, in cases where only the WRS in and near the weld are of great importance, this could be neglected. The effect of thermal treatment of welding seems to erase this effect near the weld as very good agreement is met for the measurements closest to the weld toe. Moreover, in the latter case the significantly localized effect of shot blasts around the area of impact should be accounted for the fact that the preciseness is not influenced by the remaining compressive stresses away from the weld. In the present study, where the WRS on the HFMI treatment area i.e. the weld toe are needed, this influence can be neglected.
- T_{ref} of the elements in the FZ should be equal to the melting temperature of the investigated material applied and should remain so after cooling.
- Modelling approach of boundary conditions is not as predominant for the transverse WRS, as in the case of butt welds. This was at some point expected, as the heat input and the shrinkage of the plate takes place on the bottom and top of the plate outside the plate plane. Thus, the effect would be more mediocre than in the case of butt welds were both the restrains and the heat input causing the dilatation are coplanar. The effect on the longitudinal WRS was in any case negligible.
- In order to minimize WRS and increase fatigue strength of a fillet weld with transversal stiffeners, the opposite side from the currently welded pass should be clamped down. For practical reasons, the side opposite to the cold-welded 1st pass should be restrained during welding, as in this case the effect is predominant. If preheating is applied, this approach is expected to have limited effect.
- The present engineering approach provides sufficient results for high strength steels as well.
- The proposed values for the weld metal arc efficiency found in literature were insufficient for the present analysis. Significant arbitrary increase of thermal heat input was necessary in order to calibrate the thermal analysis

and match the respective measurement results. This could be attributed to the variety of the investigated materials.

- Symmetrical welding, i.e. simultaneous welding of the top two welding passes and subsequent simultaneous welding of the bottom two passes or vice versa, can significantly decrease the peak WRS at all weld toes and lead to an improved fatigue behaviour of the weldments.
- An approach in order to simulate shot blasting by applying thermal loading instead of modelling the real mechanical effect was proposed. Although the results of preliminary analyses were satisfying, when the proposed approach was incorporated to the present weld simulation model, no significant improvement was observed. The proposed method though, seems promising as it is numerically efficient and worked well for the smaller components. A further investigation of the method was excluded as it was considered to lie outside the boundaries of the present study.
- Influence of cutting single fatigue test specimens was investigated, validating the common engineering knowledge that stress relaxation of the longitudinal WRS takes place. The results were qualitatively identical with earlier investigations on other specimens, but as quantitative deviation was observed, further investigations are proposed for this subject.

To summarize, sufficient results were met in most cases regarding the welding simulation of the fillet welds based on the precision requirements set at the beginning of the present study. The calculated WRS were considered suitable to be applied as input for the subsequent HFMI simulation.

4 Drop Tests for the Calibration of HFMI Simulation

4.1 Work Hypothesis

A series of drop tests was implemented in the framework of the present doctoral dissertation in order to allow the reproduction of a HFMI impact under monitored conditions in the laboratory and therewith, enable the characterization of the investigated material's mechanical behaviour by considering explicitly possible irregularities concerning the present deformation modus. During the initial definition of the objectives, it was considered that the applied experimental procedure should be straightforward, reproducible and therefore would not require complex experimental equipment. Main goal was the determination of the investigated material's dynamic yield strength for various strain rates inside the spectrum of interest, so that a calibration of the applied material model for the subsequent HFMI simulation could be enabled.

4.2 Methodology

During the drop tests, the HFMI pin should carry out a free fall and impact on a specimen of the investigated material with a known velocity, carrying a known mass. Impact velocity would be both analytically calculated and measured during the experiment. Analytical calculations based on Equations (31) to (38) would allow the calculation of the dynamic yield strength based on measurements of either the impact crater or the restitution coefficient e respectively. During a FE analysis of the impact using as input the known mass and impact velocity, the plastic strain rate would be defined. The obtained dynamic yield stress for the given strain rate would be initially validated through comparison with respective results from material testing under high strain rates, which was carried out in the framework of previous studies, found in literature (see [46], [47], [48], [86] etc.). The strain rate dependent material model calibrated with the results of the drop tests would be introduced to the simulation of HFMI. Comparison of this late FE model with measured profiles of residual stresses would act as a final validation step. The calibration of the material model and the upon-based FE simulation are presented in chapter 5 with the rest of the HFMI numerical investigations. A synopsis of the working steps is provided by the flowchart of Figure 89.

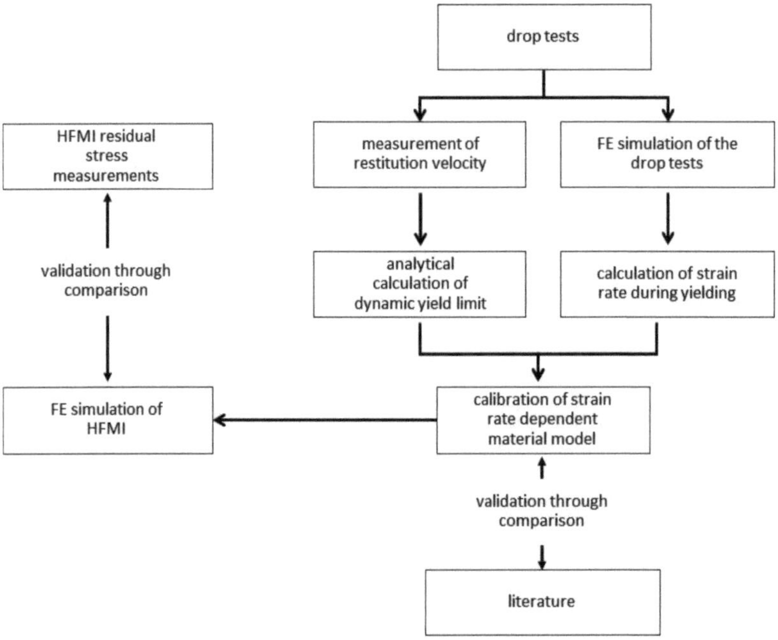

Figure 89: Flowchart for the calibration of the strain rate dependent material behaviour through the implementation of the drop tests

4.3 Investigations

4.3.1 Experimental Setup

The experimental setup of Figure 90 was constructed in order to fulfil the above-described required conditions for the drop tests. A bearing structure, consisting of four columns connected with transverse short beams all made of wood was built. Four wooden beams at the bottom along with the bottom side of the columns formed the foundation of the structure. Austenitic steel bolts and L profile connectors were applied at the joints. Steel rails were adjusted on the inner sides of the columns and a wooden cross, which carried steel wheels on each of his side, was adjusted on them. Mounting two wheels per side of the cross, one above each other so that they would both run simultaneously along the trail, would exclude excessive rotation of the cross relative to its horizontal axis and in extension a non-vertical impact of the pin. A tolerance of 5 mm was selected for each side of the cross, between the wheels middle point and the tip of the rails' cross section, so that the friction between them would be minimized, but at the same time derailing of the wheels would not be possible. A negligible rotation of ± 0.5° was allowed. Therewith, it could be considered that the cross carries out a free, vertical fall. On the bottom side of the cross, a cylindrical sinker made of plain carbon steel was adjusted in order to increase the impacting mass and in extension the contact force. Three different sinkers, with mass of 3.06 kg, 9.42 kg and 14.76 kg respectively, were manufactured and could be exchanged. Finally, a HFMI pin was mounted on the bottom of the sinker. Therewith, an impact assembly, which could land on an appropriate specimen at the end of its free fall and reproduce a single impact of the HFMI treatment was manufactured.

Plates of parent material with dimensions 100 mm x 100 mm x 10 mm like the one of Figure 91 were tested. The upper surface of the specimens, where the impact would take place, was polished prior to testing so that even small craters would become easily apparent. During the tests the impact assembly was pulled up to the desired drop height with a rope and was then left to drop. It would run along the rails until the HFMI pin would hit the investigated specimen's surface. A video-camera placed at the same plain with the specimen's surface recorded the implemented impacts. Placing a ruler at a known distance behind the impact point enabled the measurement of the rebound distance and in extension the rebound velocity of the impact assembly, as the mass of the impact assembly was known. Four impacts on a specimen of each investigated material were carried out in a sufficient distance from each other so that any interaction was excluded (Figure 91).

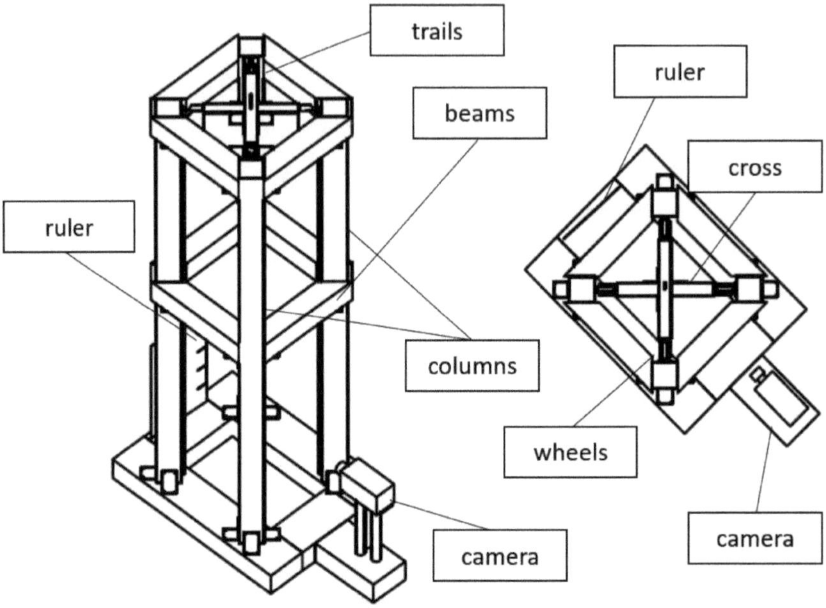

Figure 90: Experimental setup for the implementation of drop tests

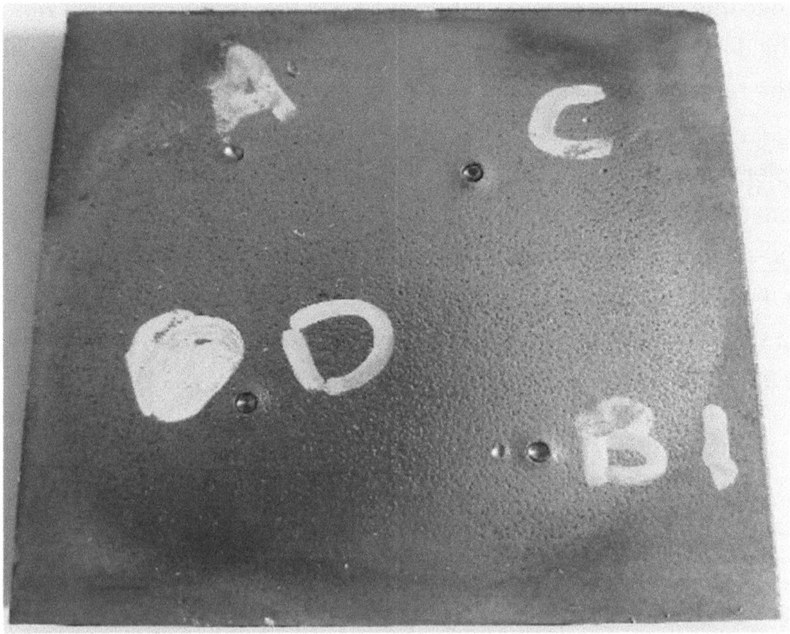

Figure 91: Upper surface of investigated specimen of S355 and the craters for the four impacts

4.3.2 Estimation of Impact Velocity

Impact velocity was calculated analytically based on trivial physics and was initially expected to be measured as well from the recorded video. During the analytical calculation, a comparison of the results by taking into consideration or neglecting the air drag shown that this influence was negligible (0.001 m/s). A drag coefficient for a perfect cylinder (sinker) ignoring the influence of the rest of the impact assembly was considered. This negligibility during the drop meant of course that the drag was negligible during the rebound too, as the rebound velocity is always lower. Drag was therefore neglected in the calculation of the restitution coefficient. In most cases, a validation of the impact velocity during the drop based on the video frames was against initial expectations not possible, as it was too high to be caught by the present video recording rate.

4.3.3 Strain Rate Calculation through FE Analysis

A finite element analysis of the drop tests was carried out in order to calculate the plastic strain rate. A strain rate independent non-linear material behaviour was applied. Considering the strain rate dependency of the yield strength and the hardening behaviour would be naturally closer to physical reality, but this would create an infinite loop to the flowchart of Figure 89. Overrunning this obstacle through an optimization process could be possible, but this would exceed the limits of the present study and it was therefore excluded. The FE analysis was carried out based on the previously presented theoretical background and the velocity-based modelling technique applied for HFMI, described thoroughly in the following chapter 5. For this reason, no further information is given on the FE model at this point. Sole difference between the simulations of the drop tests and the HFMI treatment, apart from the singularity of the impact and the strain rate independent material modelling, is the additional mass of the impact assembly. The additional mass was considered to be concentrated at the centroid of the pin, a valid assumption for a vertical impact with negligible rotations. Additionally, the singularity of the impact in these cases allowed the use of double symmetry at the impact point at the intersection of two orthogonal planes, whose intersecting line is parallel to the normal of the impact (see Figure 92).

During the initiation of contact, unexpectedly high equivalent von Mises and shear strain rates ($\dot{\varepsilon}_{\varepsilon q} > 10^5\ s^{-1}$) were evident on the contact surface for a very short time ($dt < 10^{-4}$ s). This effect was present in the case of the subsequently presented HFMI analysis and is attributed to numerical singularities, which arise due to the applied contact model. No significant influence from these singularities on the numerical results was evident, as it is described later in chapter 5. Still, the plastic strain rate was not evaluated directly after yielding, but at the point that a plastic spherical zone was created underneath the contact surface, as it is presented in Figure 92. At this point, the average of the maximum shear strain rate along the vertical line at the intersection of the two symmetry points in depth direction (line AB in Figure 92b) was extracted and was the applied strain rate for subsequent material model calibration. Exemplary, a correlation of the impact velocities and the respective calculated strain rate for S355 is illustrated in Figure 93.

Unfortunately, there is no previous respective correlation of a spherical indenter's impact speed with the plastic strain rate known to the author, which could act as reference for the present results. Still, in the above-mentioned study by Cadoni et al. [20], during Split-Hopkinson bar tests impact velocities of 9 m/s to 27 m/s were correlated with strain rates of 900 s^{-1} to 7000 s^{-1}. Yet, contact between wider circular normal surfaces (diameter of 5 mm) takes place in the former case. On the contrary, in the case of the spherical indenter the contact initiates from a point and then is applied only on a significantly smaller circular area (< 0.5 mm for the present case). Therefore, higher strain rates can be expected in the latter one due to higher stress concentration. The present calculated strain rates, which are in the same order of magnitude with those of Cadoni et al. [20] but slightly higher and for lower impact values, are thus considered valid.

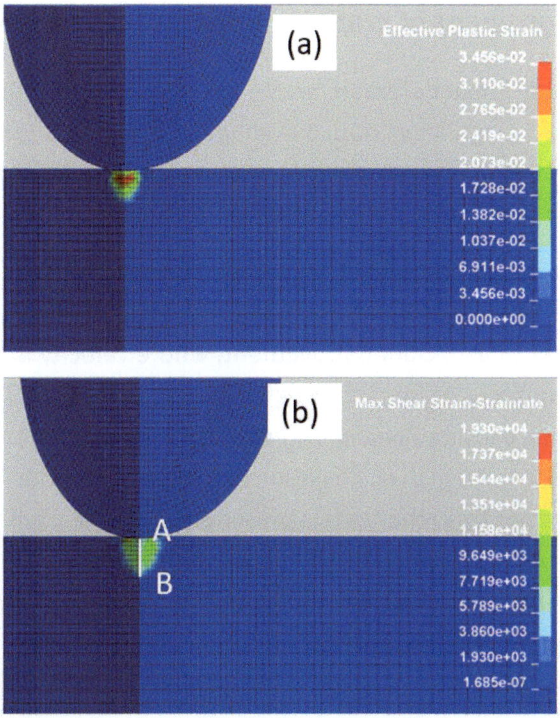

Figure 92: Introduction of the plastic strain spherical zone underneath the impact surface: a) Maximum von Mises strain rate; (b) Maximum shear strain rate

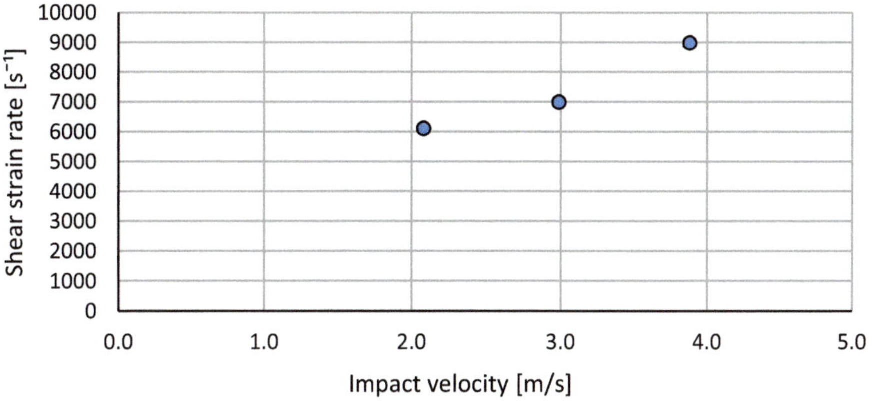

Figure 93: Average plastic shear strain rate as a function of impact velocity for the drop tests on S355

4.3.4 Measurement of crater and restitution coefficient

Dimensions of the impact's crater were measured with common laboratory equipment (graded magnifying loop, dial gauge). Exemplary the width of a pair of craters was measured under a stereomicroscope (Figure 94). Similar results were produced in both cases with a precision of ± 10 µm.

An action camera with a recording speed of 180 fps was applied for the measurement of the restitution coefficient. For this video recording speed the impact assembly at the peak of its rebound movement, where velocity is near zero, would travel between consecutive impacts for an assumed velocity of 0.1 m/s a distance smaller than 10 µm, causing a negligible error at the measurement of the rebound distance and in extension in the calculation of the rebound velocity. An exemplary set of selected frames from the tests is presented in Figure 95. Based on the principles of trivial, Euclidean geometry the real distance was calculated based on the frames taken by the camera and the measured horizontal distances between the camera, the ruler and the specimen.

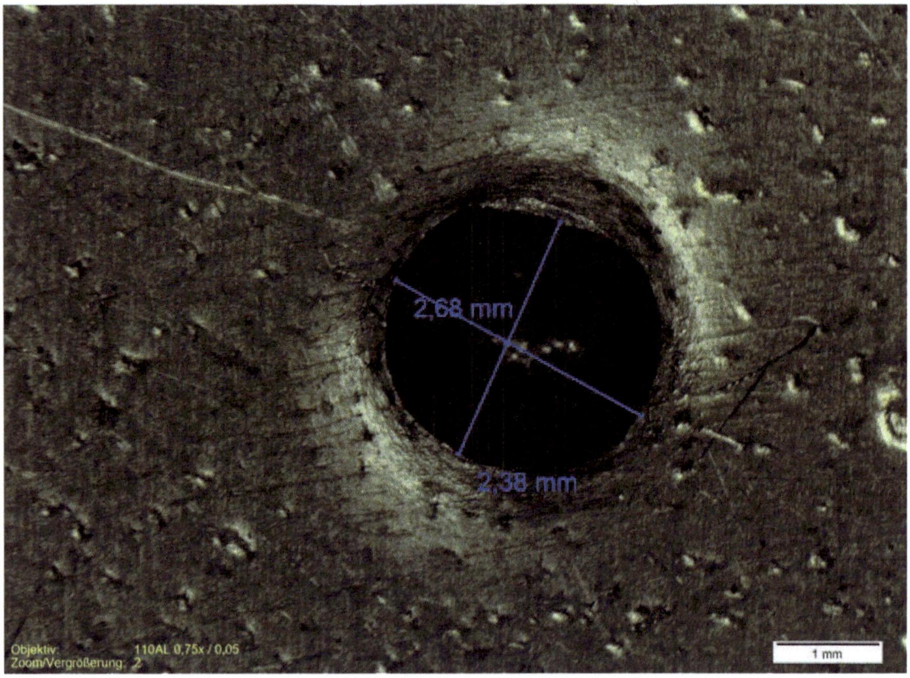

Figure 94: Crater on specimen of S960 under stereomicroscope, impact speed of 3.9 m/s

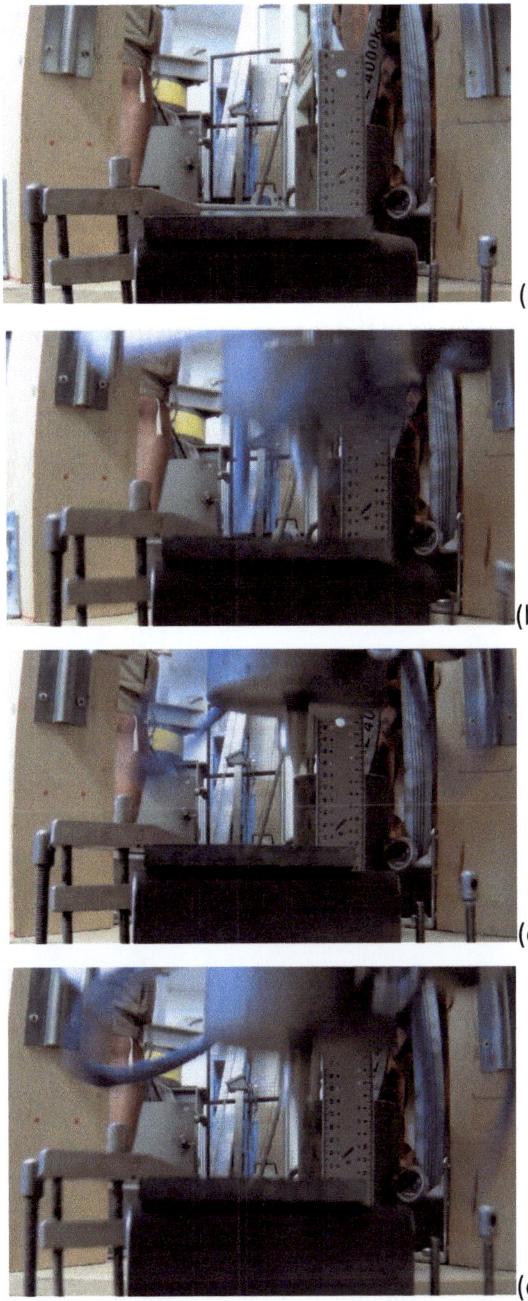

(a)

(b)

(c)

(d)

Figure 95: Selected frames from impact 3A: a) Impact complex still not in the frame; b) Impact assembly enters the frame and reaches for the target; c) Rebound peak; d) Initiation of secondary free fall

4.3.5 Analytical Estimation of the Dynamic Yield Limit

The dynamic yield strength was calculated according to the initial plan, based on the measurement of either the impact crater with use of equations (31) - (36), or the restitution coefficient with the use of equation (38).

Analytical Estimation based on the Dimensions of the Crater

Calculating the dynamic yield stress by measuring the crater's size is based on the above presented equations and the reasoning path that Lim and Stronge followed for the solution of a rigid cylinder impacting on an elastic-plastic half space [110]. Analogously to Equation (34), in the case of a spherical indenter contact-half width at yielding is correlated with the half width of the crater according to the following equation

$$a_c^3\left(\frac{1}{R}-\frac{1}{R'}\right)=\frac{a_y^3}{R}. \tag{40}$$

Contact force at yielding is calculated based on Equations (32), (33) as follows

$$F_Y=\frac{4\cdot E\cdot a_y^3}{3\cdot R\cdot(1-v^2)}, \tag{41}$$

while stress at yielding is correlated with the contact force based on the assumption of Equation (35) as follows

$$\sigma_Y=0.351\cdot\left(\frac{E}{R\cdot(1-v^2)}\right)^{\frac{2}{3}}\cdot F_Y{}^{\frac{1}{3}}. \tag{42}$$

Nonetheless, this reasoning course is based on the assumptions that were stated by Lime and Stronge [110] for the cylindrical indenter. In the present case though, significant pile-up could be observed at the periphery of the crater. This led to the introduction of a residual crater with curvature *R'* which in most cases was smaller than the initial curvature prior to unloading. Therewith, the requirements for the application of a solution analogous to that of Lime and Stronge [110] were no more valid. As expected, the present method yielded no rational results and therefore, the respective calculated dynamic yield strengths are not presented.

Exemplary, the results of the measurement of the crater and the calculation of the half contact width at yielding for S355 are given in Table 40.

Table 40: Results of the drop tests – measurement of the craters' dimensions

Impact	Impact velocity u [m/s]	Residual indentation depth δ [mm]	Half width α [mm]	Curvature after unloading R' [mm]	Indentation depth at yielding α_Y [mm]
3A	2.1	0.82	1.45	1.28	-
3B	3.0	0.69	1.60	1.86	0.08
3C	3.9	1.28	1.65	1.06	-
3D	3.9	1.34	1.63	0.96	-

It is becoming easily apparent from the above and the comparison of R' with R (1.5 mm) that the above-mentioned requirements are not met for the present impact speeds.

Finally, even for the cases, whereby R' is larger than R, like for the case 3B, the indentation at yielding δ_Y is calculated to be in the order of magnitude of decades of micrometres. More specifically, in the case of 3B it is equal to 0.08 mm or $0.05 \cdot R$. Nonetheless, in a previous study of a spherical indentation by Yan et al. [180] it was shown that yielding initiates for an indenter's radius of 0.5 mm after an indentation of just 49 nm. This later incompatibility further shows the unsuitability of the calculation of yield limit based on the dimensions of the impact's crater for the present impact velocities and for a spherical indenter.

Calculation based on Restitution Coefficient

The restitution coefficient e was measured based on the video recording of the conducted impacts. In all cases, the coefficient ranged between 35 % and 50 % validating the theoretical values proposed by Johnson for the present impact velocities [82]. Equation (38), was proposed for the case of impacting perfect spheres. As in the present case additional mass is added to the impacting pin due to the impact assembly, the equation was transformed as follows, so that the mass would be excluded from the equation:

$$e = 3.46 \cdot {\sigma'_Y}^{\frac{5}{8}} \cdot {E^*}^{-\frac{1}{2}} \cdot {V'}^{-\frac{1}{4}} \cdot \rho^{-\frac{1}{8}}, \tag{43}$$

which based on the assumption for the rigidness of the pin becomes

$$e = 3.46 \cdot {\sigma'_Y}^{\frac{5}{8}} \cdot \left(\frac{1-v^2}{E}\right)^{\frac{1}{2}} \cdot {V'}^{-\frac{1}{4}} \cdot \rho^{-\frac{1}{8}}, \tag{44}$$

which in turn, for $v = 0.3$ and $\rho = 7850\ \mathrm{kg} / \mathrm{m}^3$, becomes

$$\sigma'_Y = 0.90 \cdot e \cdot E^{\frac{1}{2}} \cdot {V'}^{\frac{1}{4}}. \tag{45}$$

As already mentioned above, Equation (38) was proposed by Johnson based on the premise that the average pressure during full plastification is equal to 3 times the yield stress ($p_m{}^p \approx 3 \cdot \sigma_Y'$). The sensitivity of this assumption on the yield strength has as well already been described and thus, a modification taking into consideration the changed yield stress in the case of the two high strength steels had to be adopted. Interpolating the values from Jackson and Green [81], Equation (45) becomes for S690

$$\sigma'_Y = 1.06 \cdot e \cdot E^{\frac{1}{2}} \cdot {V'}^{\frac{1}{4}}, \tag{46}$$

and for S960

$$\sigma'_Y = 1.10 \cdot e \cdot E^{\frac{1}{2}} \cdot {V'}^{\frac{1}{4}}, \tag{47}$$

respectively. Evaluation of the dynamic yield stress based on these equations was carried out for the three investigated materials. In all cases, the measured coefficient of restitution was increased by 10 %, which are the losses due to elastic wave propagation during a spherical indentation, according to [110].

Equation 44 yielded reasonable results for S355. Nonetheless, in the case of the two high strength steels, apart from one case, a dynamic yield lower than the static one was calculated. Therefore, the above-mentioned reasoning course cannot describe phenomena, which correlate with the dynamic behaviour of these high strength steels in a deformation case like the present one. Consequently the results regarding the two high strength steels are excluded from the rest of this evaluation. The calculated yield stresses for S355 are presented in Table 41, along with the analytically calculated impact speed and the respective strain rate, received by the FE analysis of the drop tests.

The calculated dynamic limit is compared with the results of the previous studies for the same material in the diagram of Figure 96. The results of the drop tests lie much closer to those of Cadoni [20] for the case of compressive loading. Strain rate-dependency in the case of tensile loading becomes predominant at strain rates of around 100 s^{-1}, while for the compressive tests of Cadoni and the present drop tests significant increase of yield stress is met for strain rates higher than 5000 s^{-1}. Therewith, the above-stated assumption that the compressive strain rate sensitivity of the material is predominant for the case of spherical indentation, and in extension of the HFMI treatment, is validated. More specifically, the material is expected to be less strain rate sensitive during the HFMI treatment than the material behaviour, which was extracted from tensile tests for same strain rates.

Table 41: Results of the drop tests: calculation of the dynamic yield stress based on restitution coefficient

Impact	v [m/s]	$\dot{\varepsilon}$ [s^{-1}]	σ_Y' [N/mm²]
3A	2.08	6120	647
3B	2.99	7000	607
3C	3.88	8990	618
3D	3.88	8990	547

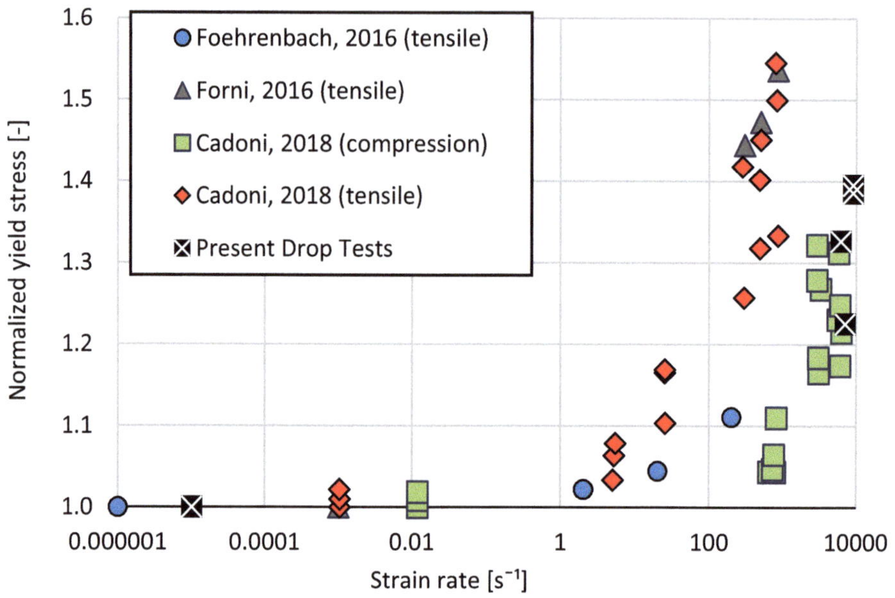

Figure 96: The dynamic yield strength of S355 as a function of the strain rate, results from present drop tests and previous studies by Foehrenbach et al. [44], Cadoni et al. [20] and Forni et al. [47][1]

4.4 Summary and Open Questions

A series of drop tests with a spherical indenter, identical to the HFMI pin, was carried out on plates of the investigated materials, namely S355, S960 and S960, in order to estimate their dynamic yield limit. Initially, two different estimation methodologies were planned, based on measurement at different strain rates of either the crater's dimensions or the impact's coefficient of restitution. The former method did not produce reasonable results, as the requirements for the application of the respective analytical equations were not fulfilled. The latter yielded satisfying results, which confirmed initial expectations and showed good agreement with previous respective results regarding the strain rate dependent, compressive yield behaviour of S355. Nonetheless, even by this second estimation method, no reasonable results were met for the high strength steels. It is assumed that certain phenomena of the dynamic behaviour of these high strength steels cannot be captured by the present analytical solution, even

[1] the static value for the present drop test series (at 0.000002 s^{-1}) is the one acquired by the respective static tensile test on specimens of the same material batch and is given as a reference

though an adjustment was proposed based on data from previous studies. In all cases, the strain rate was calculated by means of FE analyses for strain rate independent yielding. As a last-step-validation of the calculated material behaviour, a strain rate dependent material model was calibrated and introduced to a FE model for the HFMI treatment of a specimen of parent material. This analysis is presented in the next chapter of this dissertation along with the rest of the numerical investigations regarding HFMI. The applied approach shows potential for calibrating the material behaviour of steel for the simulation of HFMI taking into consideration the irregularities of material behaviour during spherical indentation. Still, many difficulties arise regarding different aspects of the present solution, which lead to discrepancies, especially for the high strength steels.

In order to eliminate the above-mentioned difficulties and to further develop the present approach, the implementation of the following steps in future work, as they exceed the limits of the present study, is encouraged:

- More tests and for a wider variety of impact speeds could offer a better overview regarding the distinctiveness of the steels with higher yield point. Exclusion of outliers and a better adjustment of the analytical equations could be enabled by a larger sample population.
- Further validation is possible through the measurement of the residual stresses introduced during the drop tests on the investigated plates by X-ray or another measurement method. The simulation of the RS field by a FE analysis of the drop tests, this time by considering the calibrated strain rate dependent material model, could enable a more robust validation.
- The influence of the specimens' thickness could as well be investigated. All the implemented analytical calculations were based on the premise that the deformed plate is behaving like an infinite half-space. Although in the present case the thickness of the specimen was considered sufficiently larger than that of the crater's depth (larger than 6.5 times in all cases) some effect could still be evident. Investigation of thicker and thinner plates is encouraged.

Finally, it has to be underlined that the material model, which was calibrated by the present drop test, was subsequently introduced to a simulation of specimen of parent material. In order to calibrate a model for the simulation of HFMI-treated weldments, specimens whose microstructure is similar to that of the weldments' HAZ should be tested. This could be enabled through an appropriate thermal pre-treatment of the specimens, for instance with a Gleeble® thermo-mechanical simulator instrument.

5 FE Simulation of HFMI

5.1 Methodology

A bottom up approach was adopted for the validation of the HFMI modelling methodology, as in the former case of welding simulation. Initially a simpler, unnotched component of parent material was simulated. The results of this preliminary model were compared with previous numerical studies and RS measurements, which were found in literature [44], [145]. The validated model was then applied for evaluating the influence of different aspects of modelling and the treatment on the calculated RS. Conclusions regarding the scaling effect, boundary conditions and modelling of material behaviour were drawn. These conclusions were valuable for the subsequent series of simulation, wherein the fillet welds from the research project *HFH-Simulation* [146] were investigated, in order to fulfil the goal of the present study. RS measurements from *HFH-Simulation* were used for a further and final validation of the established approach. The influence of WRS from the preceding weld simulation was considered as well. After validation of this model, further aspects of the treatment were investigated.

Commercial FE software *LS-DYNA* [113] was applied in all HFMI simulations of the present study. Explicit dynamic FE analysis was carried out and respective mesh of 8-node solid elements, *hexaedra*, was applied on the simulated components. More details regarding the FE theoretical background is given in Appendix B. As described above, proper modelling of the HFMI pin movement and material behaviour is decisive for an accurate simulation. In all investigations of the present study the HFMI pin was simulated as a rigid body due to its negligible deformation and significantly higher hardness in comparison to that of the treated surface. Two different diameters were considered, 3 mm and 4 mm, corresponding to the pin size of the *HiFIT* and *PITec* devices.

Two different approaches were taken into consideration for simulating the movement of the HFMI pin. When a displacement-based approach is considered, mass and inertia of the pin can be selected arbitrary, as they have no effect on the calculated RS; a rigid volume is pressed upon the treatment surface with a prescribed motion similarly to a punching process. On the contrary, when a force- or velocity-based[1] approach is adopted, the mass and the inertia of the real pin has to be considered; an arbitrary accelerating force is applied on the pin, calibrated on measurements on the real device, and the resulting impact velocity is dependent on these properties. When the pin hits the treatment surface, it has initial velocity and no further restrain on its vertical axis movement. The horizontal movement of the pin, along the treatment line, was modelled stepwise (displacement-based) for all simulations as this has no influence on the impact and in extension to the modelled RS.

Coulomb's frictional contact (Eq. (28)) between the HFMI pin and the treated surface is taken into consideration in all cases. Friction coefficient μ is calculated automatically by the applied Software *LS-DYNA* through interpolation between a static (μ_s) and a dynamic value (μ_d) according to the following Equation (48):

$$\mu = \mu_d + (\mu_s - \mu_d) \cdot e^{-c|v_{sliding}|}. \tag{48}$$

The values of 0.78 and 0.42, which were initially applied respectively for the two coefficients, were found in [6]. Several values can be found in literature for these coefficients. Although no significant influence on the introduced RS due to their variation was expected according to initial expectations, further lower values were tested, as it is described below. A damping model according to Equation (27) with a factor ξ initially set equal to 0.5 was applied as in previous numerical investigations of impacts ([89], [90] etc.). Flanagan-Belytschko hourglass control is applied in all cases for the elements inside and near the treatment area [43]. Two different types of this hourglass control, the classical and the exact volume integration for solid elements (see [114]), were compared with no significant deviation, with the latter one though being the most stable numerically. It was therefore adopted for all subsequent analyses.

Suitable modelling of material behaviour is predominant for the precise simulation of HFMI, a prerequisite that becomes clear even from a preliminary literature

[1] the velocity-based term is preferred in the present case as the pin is accelerated through the application of an arbitrary force until it reaches the required impact velocity, the method though is physically equivalent to a force-based, or an acceleration-based approach

review. According to initial expectations, strain rate dependency had to be taken into consideration otherwise erroneous results might arise. The Cowper-Symonds model of Equation (12) combined with a bilinear σ-ε material behaviour with kinematic hardening is applied in all subsequent investigations of the present study for the consideration of viscous material behaviour.

It should be highlighted at this point that during the displacement-based simulations, questions regarding the application of a viscous material behaviour arise. Although a displacement-based simulation has been carried out in previous studies combined with strain rate dependent material behaviour ([44], [108] etc.) this approach should be applied cautiously. Defining an arbitrary vertical, over time displacement-based motion of the pin is straightforward from the modeller's perspective but directly influences the strain rate, under which the treated surface is deformed. Hence, yielding and the introduced RS depend on this arbitrary defined displacement. Unless measurements of this displacement over time during a real treatment are made, which would require a very elaborate experimental investigation, this modelling approach is invalid. If the contact duration is known, the assumption of a sinusoidal displacement of the pin inside this contact time could be valid. Nevertheless, if the contact force and in extension the impact velocity are known, applying a velocity-based approach for simulations considering a viscous material behaviour should be preferred instead as such an approach is more valid and closer to physical reality.

5.2 Convergence Analysis

Preliminary investigations considering a single impact of a HFMI pin with a diameter of 3 mm on a component of parent material made of S355 were carried out in order to define the maximum mesh dimensions at the contact area, which would produce mesh-independent results and enable at the same time a computational efficient simulation. A displacement-based vertical movement of the HFMI pin of 0.5 mm was considered, as in this case comparing the influence of different mesh sizes was the goal and not the precise simulation of the RS. The modelled HFMI groove was on purpose significantly higher than the real one, so that the selection of mesh size would be conservative. Strain rate dependent behaviour was taken into consideration as well, as variated yielding behaviour could require different mesh size. Due to consideration of a single impact in this case, adoption of double symmetry was enabled.

Four different discretization setups with element dimensions of 0.2 mm, 0.1 mm, 0.075 mm and 0.05 mm to all directions were applied. The longitudinal RS on the surface from the centre of the impact and transverse to the impact crater were selected as an index of convergence. The results of this convergence analysis are given in Figure 97. It is becoming evident that the results diverge for the two first applied mesh sizes but converge for the latter two. Therewith, a required element length to pin diameter ratio of 1/40, which was proposed for the simulation of shot peening in [77] and [90], is validated in this case as well. More information on these investigations and the range of variations are not significant but can be found in [168].

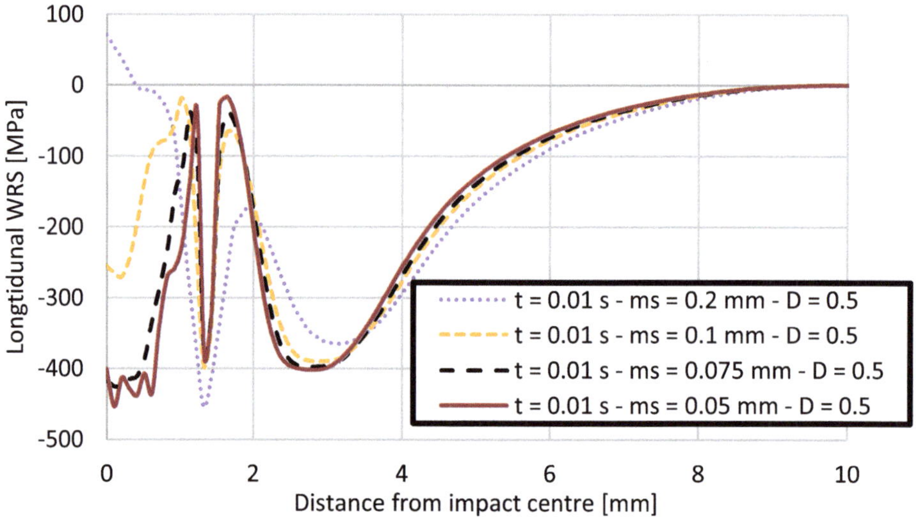

Figure 97: Convergence study for the numerical investigation of HFMI treatment – RS after 0.01 s of simulation with global damping *Ds* = 0.5 (*Ds* and mesh size are marked as *D* and *ms* rspectively)

5.3 Component of Parent material

5.3.1.1 Investigated Component

The specimen of parent material with dimensions 20 mm x 20 mm x 10 mm, which is presented in Figure 98, treated with a 4 mm diameter HFMI pin, was considered in a first-step validation of the applied modelling approach. An adaptive mesh approach was selected in this case. At the contact area and near the surface, discretization with element size of 0.125 mm or even smaller was applied, gradually coarsening in depth direction. Transverse to the treatment line and 4 mm away from it, transition mesh was adjusted, so that coarsening in this direction could be fulfilled too. The applied mesh is illustrated in Figure 99. During preliminary investigations, the treatment begun and ended 2 mm away from the respective edge. Nonetheless, as a significant influence of the edges on the introduced RS field was obvious, this distance was increased to 5 mm leading to a reduced treatment length of 10 mm. Therewith, this edge influence was successfully eliminated and this treatment setup was applied for all subsequent investigations of these specimens.

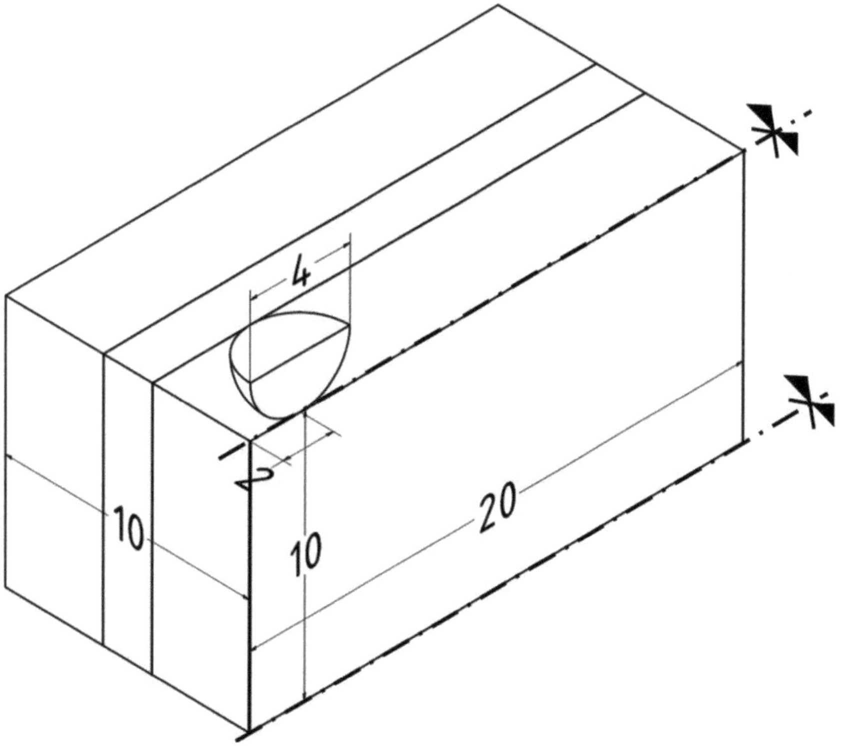

Figure 98: Investigated specimen of parent material with dimensions 20 mm x 20 mm x 10 mm with a symmetry plane at the middle – A HFMI pin with diameter of 4 mm is considered

Investigated geometry and material were selected in such a way that the results would be directly comparable to the WRS measurements carried out on identical real components in [44] and [145], in order to validate the applied approach. During these previous studies, the WRS profiles were measured in depth direction with neutron diffraction. Due to the implemented measurement technique, the measured RS relate with measurement gauge volumes of 0.5 mm x 0.5 mm x 0.5 mm in the longitudinal direction and 0.5 mm x 10 mm x 0.5 mm in the transverse one. The measurement procedure is thoroughly described in [145]. In order to enable a direct comparison, averaging of the modelled RS in the depth direction at the integration points, which are included in the respective volume of each measurement, had to be carried out. Additional measurements of the transverse RS were carried out on the upper surface of the components with X-ray diffraction [44], [145].

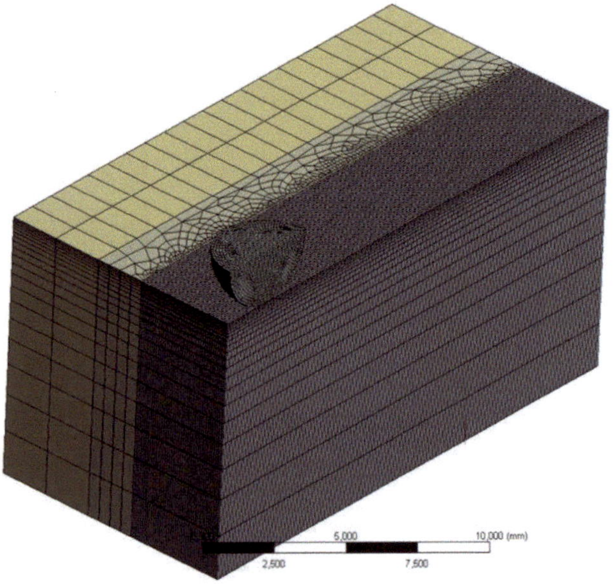

Figure 99: Mesh applied on the specimen of parent material

Bilinear material behaviour was assumed for the investigated S355, S690 and S960. It was calibrated based on experimental data from [44] and the project *HFH-Simulation* [146]. The experimentally determined σ-ε and the respective applied bilinear models are presented in Figure 100. The viscous material model Cowper-Symonds (Eq. (12)) was calibrated for S355, initially based on data from [44] and later on the drop tests, which were presented in chapter 4 along with the results provided by Cadoni et al. for high strain rate compression [20]. Calibration for the two high strength steels was carried out based on the data found in [145] and [146]. The calibrated parameters are presented in Table 42. The statistical regression results are presented in Appendix C.

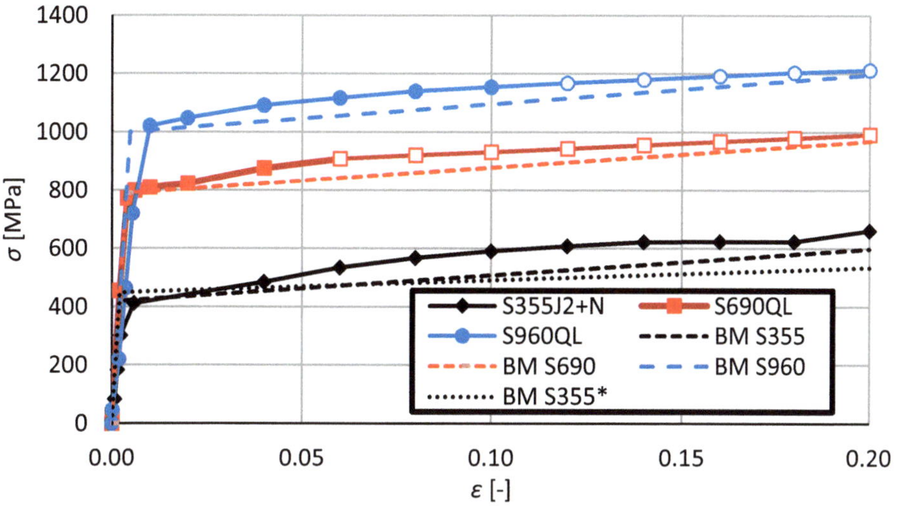

Figure 100: Diagram of σ-ε (real) and the applied bilinear material models for the investigated material – Hollow symbols mark extrapolated values – BM stands for bilinear model

Table 42: Calibration of the bilinear material behaviour and the Cowper-Symonds model

Material	σ_y [MPa]	H [GPa]	Δ [-]	γ [-]	Calibrated based on
BM_S355A	355	2.2	18250	5	Gkatzogiannis et al. [48]
BM_S355B	450	0.9	353500	5	Foehrenbach et al. [44]
BM_S355C	450	0.9	613500	4	Drop Tests and Cadoni et al. [20]
BM_S690	780	0.9	10360	3.4	*HFH-Simulation* [133] and Schubnell et al. [145]
BM_S960	1050	1	4156	4.1	*HFH-Simulation* [133] and Schubnell et al. [145]

5.3.1.2 Boundary Conditions

The real component, which was selected for the present validation, was treated with a HFMI device[2] in three consecutive passes with a moving speed of 24 cm/min and a frequency of 90 Hz [44]. This treatment setup equals a travelling step between consecutive impacts[3] of 0.044 mm. Specimens treated with a lower

[2] a PITec device was applied in [44]

[3] hereinafter called simply travelling step

travelling speed of 12 cm/min[4] have been measured as well in [44]. The influence of the differentiated, lower speed on the RS was negligible. Therefore, an analogous simplification during the numerical investigation of the treatment assuming a larger step than the real one could be adopted, as the number of modelled impacts and in extension, the travelling step directly influence the computational time. Of course, an upper limit to this simplification should hold. The maximum step between consecutive impacts, which does not affect the preciseness of the results, should be selected in order to produce a numerically more efficient simulation. Regarding the vertical movement of the pin, two approaches, a displacement- and a velocity-based, were considered in the present investigation with three consecutive passes being modelled in both cases. The measured groove depth after each treatment pass on the real component was 0.143 mm, 0. 193 mm and 0.223 mm respectively [44].

For the simulation of S355, a step of 0.4 mm between consecutive impacts was initially assumed, as it was proposed in [150] and later applied in [44], [68], and [108], in order to reduce the computational cost. In the case of the displacement-based simulation, it was initially assumed that 0.4 mm would indeed be adequate; the pin movement towards the surface is prescribed and consecutive impacts must simply overlap sufficiently with each other geometrically so that a smooth, deformed groove is achieved along the treatment line. It must be underlined that the step of 0.4 mm was proposed in [150] for displacement-based simulations. In the case of velocity-based approaches though, questions arose prior to the present series of analyses regarding the suitability of such a step size. Introducing a groove depth equal to the real one through the simulated pin movement was selected at this point as a first-step validation criterion of the model. It should be achieved before modelled and measured RS could be compared.

The displacement-based vertical motion of the pin was programmed with a constant speed of 0.288 m/s downwards and equal to the real groove of the treatment area after each pass. The elastic spring back after removal of the pin was neglected for all displacement-based simulations. The rebound speed of the pin was set much higher in order to reduce the computational time, based on the rational assumption that it has no significant influence on the calculated RS. This

---

[4] travelling speed proposed by the manufacturer

was validated by an additional identical simulation with lower rebound speed, the results of which are omitted for the sake of space. In any case, sufficient time was provided for the elastic springback of the deformed material to take place. The vertical and the longitudinal displacement of the pin over time for a 3-pass displacement-based simulation are presented in Figure 101. It is becoming evident by this diagram that the HFMI pin is displaced along the treatment line for one travelling step (longitudinal displacement of 0.4 mm in this case), when it is not in contact with the specimen (zero vertical displacement[5]). The longitudinal displacement of the pin between the consecutive impacts is completed in the model in approximately 6 ms and significantly quicker than in reality (approx. 33 ms) in order to reduce the computational time. Once again as the pin is not in contact with the specimen and as long as the above-mentioned condition regarding travelling step (impact density) is held this has rationally no influence on the results. After positioning of the pin over the next impact point, it is displaced downwards, towards the treated surface (vertical displacement of -0.2 mm during the first pass). The presented values are the initially assumed and are presented exemplary. They were changed in subsequent analyses as it is described below.

[5] the displacement orthogonal to the upper surface of the treated specimen, hereinafter called simply vertical displacement

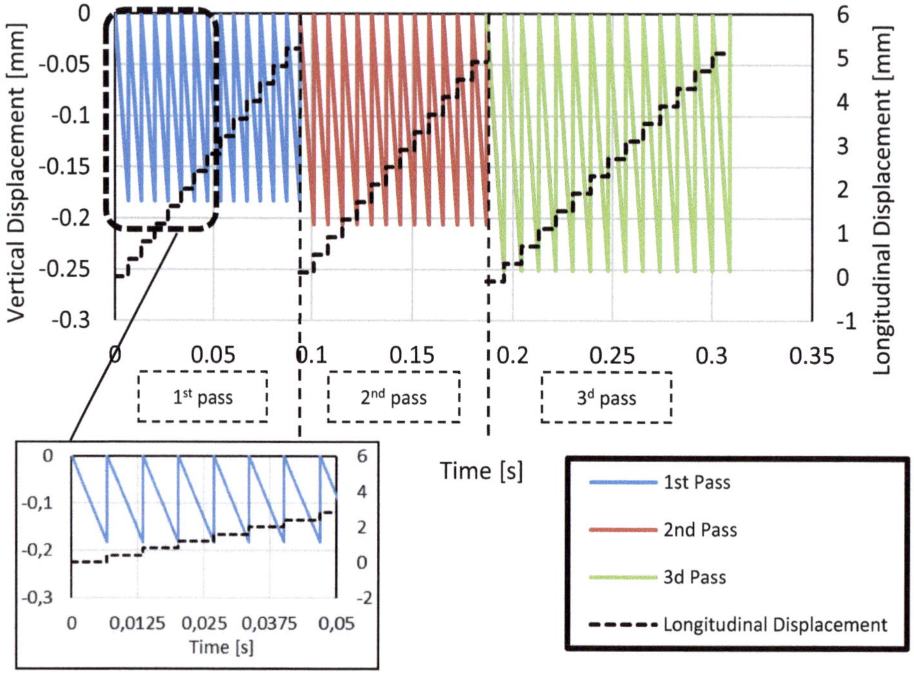

Figure 101: Diagram of vertical and longitudinal displacement of the pin, for three consecutive passes during a displacement-based simulation

Regarding the velocity-based approach, the vertical movement of the pin was calibrated through a *trial and error* procedure. The concept of the velocity-based approach is illustrated by the diagram of Figure 102. The corresponding, resulting impact velocity is presented in Figure 103. An arbitrary force (-1.1 kN), significantly high so that the computational time necessary for the acceleration of the pin would be minimized, was applied on the pin until its velocity would reach the desired value (2.7 m/s). When the pin would reach the desired impact velocity, the accelerating force was set to zero. The acceleration of the pin through this accelerating force was taking place between consecutive impacts, when the pin was not in contact with the component. At the same time, the pin would move for one step along the treatment line (0.4 mm longitudinal displacement). Hence, when the pin was hitting the treated surface, it had the desired impact velocity, no external force was applied vertical to its axis and no displacement along the treatment line would take place. Inertial properties had to be assigned to the pin, although it was modelled as a rigid body, as the energy that it transfers depends on both its velocity and mass. The mass and inertia of the real *PITec* pin were used. The calibration process was repeated until a preciseness of ± 0.1 m/s for the impact velocity in each pass was met. The preciseness of the simulated impact velocity is becoming evident from Figure 103. Peaks at the beginning of each treatment pass, i.e. in the first two impacts, were excluded from this consideration as the velocity there was unavoidably higher, but always having a rational value inside the measured range. Initially, an impact velocity of 2.7 m/s, was modelled as it was proposed in the framework of *HFH-Simulation* for the *PITec* device. The above-described acceleration and simultaneous longitudinal displacement of the pin for one travelling step between consecutive impacts is becoming evident from the following diagram (Figure 102).

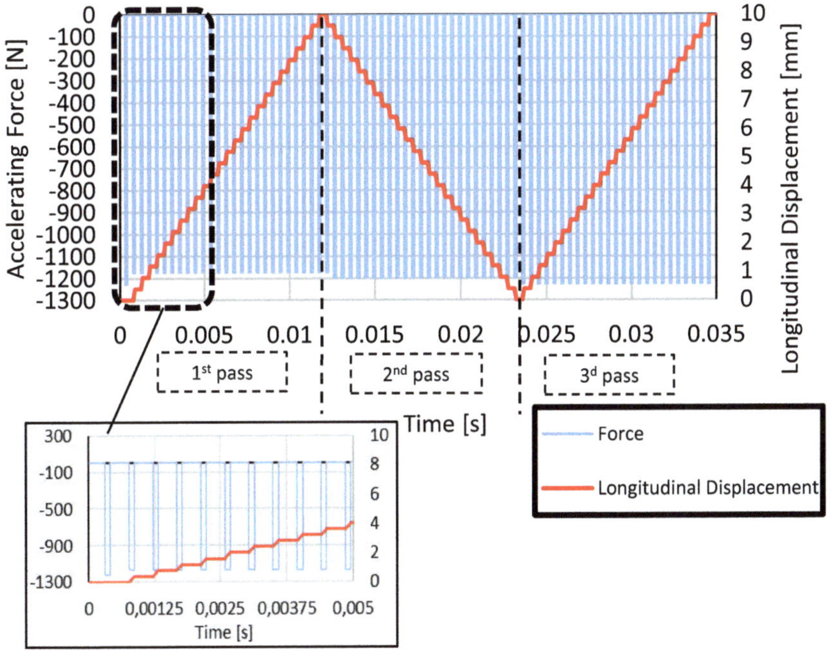

Figure 102: Diagram of accelerating force and longitudinal displacement of the pin over time, for three consecutive passes during a velocity-based simulation

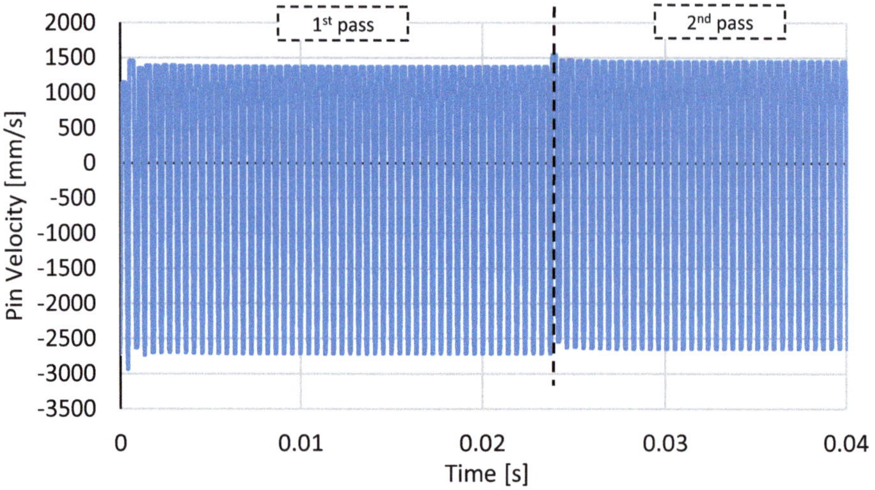

Figure 103: Diagram of vertical velocity of HFMI Pin for the first 0.04 s of the simulation

5.3.1.3 Analyses with the Displacement-based Approach

As it was described above, a displacement-based approach was initially developed and presented for the first time in [54], investigating only S355 as a first step. The vertical movement of the pin was calibrated based on the groove, which was measured on the real specimens in [168]. Material model BM_S355A was applied in two different simulations the *355_DB_RD* and *355_DB_RI*, including and excluding[6] strain rate dependency respectively. As it was mentioned above, the introduction of a viscous material behaviour in a displacement-based approach is invalid without further data and it should be carried out with caution. Nonetheless, it was applied in the present case in order to enable a first evaluation of the influence of viscosity at least qualitatively. The vertical displacement of the pin was carried out with a velocity of 2.7 m/s prior and during the contact and up to the targeted depth, so that the strain rate would be at least at same order of magnitude with the real one. When the results of this approach were published for the first time in [54], the above-mentioned averaging procedure of the modelled RS was not implemented. Instead, the results on the integration points underneath the weld toe were directly compared to the measurements. A significant deviation between measured and modelled RS was met. The averaged

[6] setting coefficients Δ and γ equal to 0

results of the displacement-based models are presented in Figure 104 along with the unaveraged ones, as they were published in [54], and the measured RS.

Both dipslacement-based approaches, *355_DB_RI and 355_DB_RD*, underestimated the longitudinal RS near the surface, while a better agreement was met in the case of the transverse RS at the same region. The overall disagreement, even qualitatively, of the rate-indendent approach *355_DB_RI* dictated the introduction of strain rate dependency to the subsequent analysis *355_DB_RD.* The rate-dependent model, which still underestimates the longitudinal RS and shows no good agremment in the first two mm below the surface, exhibits due to an overall shift of both RS profiles a significantly improved qualitative agreement with the measured RS. Therewith, the initially assumed significance of strain rate dependency for the HFMI simulation seems to be validated but further investigations were necessary at this point. Averaging of the numerical results did not improve their agreement with the measurements. A further improvement of the displacement-based approach, by assuming a sinuisoidal movement of the pin or considering the elastic springback of the treated material during calibration of the pin's movement, was not adopted at this point as it was decided to apply the more realistic velocity-based approach in the subsequent simulations.

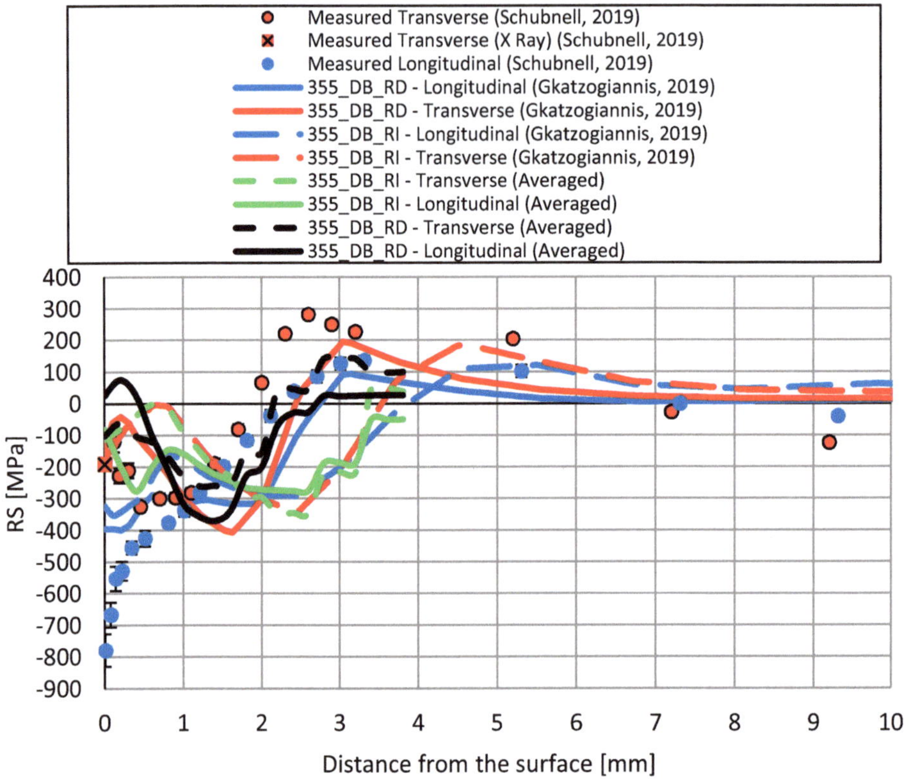

Figure 104: WRS simulated with the displacement-based method, compared with previous numerical results [54] and WRS measurements found in [44] and [145]

5.3.1.4 Analyses with the Velocity-based Approach

Calibration based on the Introduced Groove

During preliminary analyses with velocity-based models, significant conclusions were drawn regarding the step between consecutive impacts, the impact velocity and the material modelling, which would affect the set up of all subsequent analyses. A treatment with a step of 0.4 mm between consecutive impacts, an impact velocity of 2.7 m/s and for a considered material behaviour BM_S355, introduced a trace depth of just 50 % of the real one. Therewith, the unsuitability of a step of 0.4 mm for the case of velocity-based simulations is becoming evident. Simunek et al. had anyway applied this step in [150] only for displacement-based simulations, wherein the pin is "punched" into the treated material, as it was described earlier. The adoption of this increased step was practical and it was implemented in other previous studies as well [44], [68], [187], with no further investigation on its suitability though. Of course, in the present case a low impact speed could be attributed for the shallow groove as well. Given that the present value of 2.7 m/s lies in the boundaries of the real, measured impact speed and the step of 0.4 is 5 times larger than the real one, a decrease of the step to 0.2 mm was decided for subsequent analyses. This led to the introduction of a deeper trace by the model, but still lower than the real one. Groove depths equal to 79 % and 74 % of the real one were documented for the 1st and 2nd pass respectively[7].

At this point, a further decrease of the applied step was avoided as it would increase significantly the computational time. Instead, an increase of the impact velocity to 3.79 m/s, a value that as well lies inside the measured spectrum and closer to the mean of the measured real impact velocities, was preferred. Moreover, new lower values for the friction coefficients μ_s and μ_d of Equation (48), equal to 0.3 and 0.15 respectively, were selected according to common engineering knowledge. The adopted changes led to a simulated groove depth, which lies much closer to the real one for all three passes of the treatment, as it can be seen in Table 44, where the results of these preliminary analyses are summarized. An acceptable deviation smaller than ± 5 % in all cases was documented. As the considered material behaviour as well influences the introduced trace, the travelling step of 0.2 mm and the new lower values for the friction coefficients were adopted for all subsequent analyses. Prior to any evaluation of RS results, the impact velocity would be adjusted for each material taking values from inside the measured spectrum, until an acceptable agreement with the real groove would be met. If this would not be sufficient, then the travelling step would be further reduced.

[7] during the 3d pass numerical problems were met and therefore the respective results are excluded from the present evaluation

Table 43: Measured and simulated groove depth of the HFMI treatment – Travelling step in mm and impact velocity in m/s

Case	Groove depth [mm]		
	1st Pass	2nd Pass	3d Pass
measured in [44]	0.143	0.193	0.223
simulated with 0.4 mm - 2.72 m/s	0.072	0.093	0.106
simulated with 0.2 mm - 2.72 m/s	0.113	0.143	-
simulated with 0.2 mm – 3.68 m/s	0.148	0.184	0.217

Influence of Simulated Material Behaviour

Simulation of the present component for all three investigated materials S355, S690 and S960 based on material models BM_355, BM_690 and BM_960 were carried out by both considering and neglecting in all cases the strain rate dependency. In the former case of S355, an elastic - perfect plastic model behaviour was tested as well[8]. The through-depth, averaged longitudinal and transverse RS for S355 at the centre of the HFMI groove at the middle of the component are given in Figure 105 and Figure 106. Figure 107 and Figure 108 show the respective results for S690. Finally, Figure 109 and Figure 110 illustrate the respective transverse and longitudinal profiles regarding S960. In all cases, the respective measurements found in [145] are presented along the present simulation results. Small fluctuations, which are met in some simulated profiles, are attributed to the averaging process of the results, without this having a significant macroscopic influence either qualitatively or on the magnitude.

Regarding the RS of S355, all three investigated models provided qualitatively similar profiles. Both longitudinal and transverse RS profiles are compressive near the surface with a transition to the tensile area in a depth of approximately 1.5 mm to 2 mm. Regarding the transverse RS, very good agreement with the measured profile is met on most of the measurement points in all cases apart from the compressive strain rate dependent model. For all models, the largest deviation from the measured RS is met at a depth of approximately 0.75 mm, where the

[8] setting $H = 0$ for the BM_355

peak of both the measured and simulated compressive stresses is located. This deviation is slightly higher for the strain rate dependent simulation, although the difference can be considered negligible. On the contrary, longitudinal simulated RS profiles present a significant quantitative deviation with each other. The tensile strain rate dependent model estimated with very good agreement the measured RS on and near the surface, but underestimated them on a depth larger than 1 mm. On the other hand, both strain rate independent material models underestimated the RS near the surface. The compressive strain rate dependent model produces a similar profile with the tensile one, overestimating though overall the RS.

The significantly better agreement of the tensile strain rate dependent material model for the RS near the surface, where fatigue cracks initiate, reveals the predominance of the strain rate dependency for the present simulation, verifying in this first-step analysis the respective initial assumption regarding its importance. When the material exhibits significant viscosity, the layers near the surface where the high strain rates are met during the impact, yield at higher stresses absorbing more kinetic energy and therefore, the peak of the WRS is shifted nearer to the surface. Therefore, the introduction of viscosity becomes necessary even for a qualitatively precise simulation.

On the contrary, the higher preciseness of the tensile versus the compressive strain rate dependent material model, does not overthrow the assumption regarding the predominance of the compressive behaviour in the case of HFMI yet. The better agreement of the tensile model can be attributed to the fact that it was calibrated based on the same batch of the investigated material, applied in [44] as well. On the other hand, during the calibration of the compressive strain rate dependent material model, data from [20] for the same material but another batch was considered along with the results of the drop tests in order to increase the statistical sample. As it was discussed in [59] though, the strain rate sensitivity of the metallic alloys can differ significantly from a batch of a material to another. Therefore, no assumption regarding the predominance of the compressive behaviour could be made, unless more measured and simulated results from the same batch could be compared.

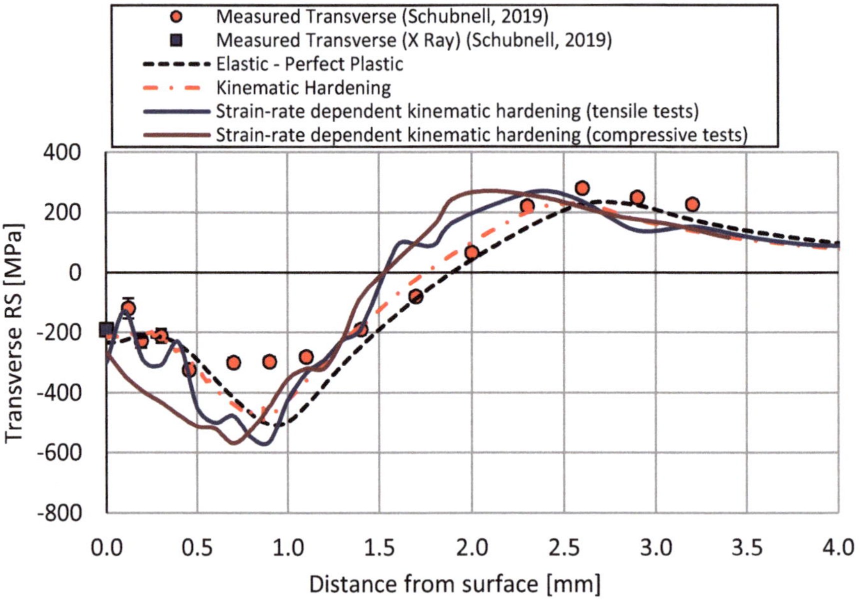

Figure 105: Simulated and measured transverse RS - Component of parent material S355 – (Schubnell, 2019) measurements were found in [44] and [145]

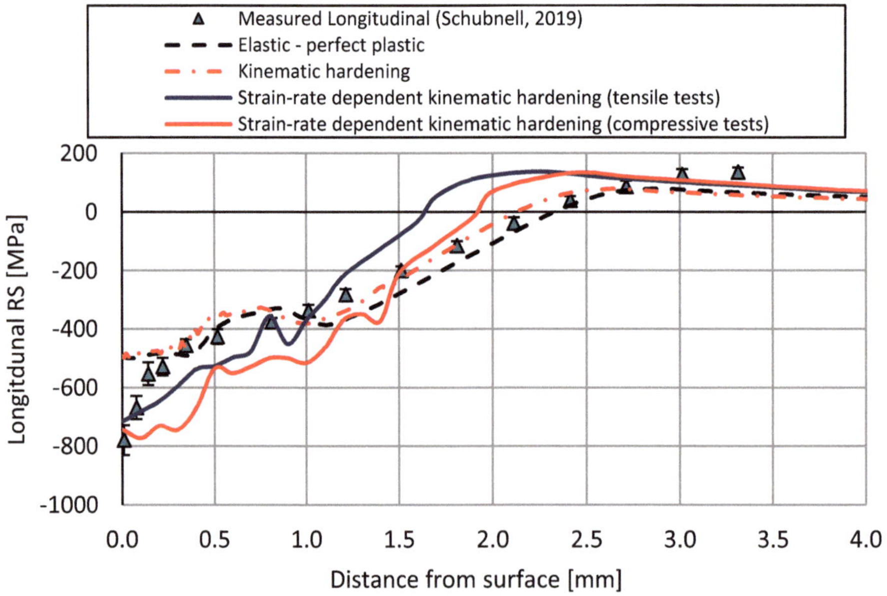

Figure 106: Simulated and measured longitudinal RS - Component of parent material S355 – (Schubnell, 2019) measurements were found in [44] and [145]

Qualitatively similar RS profiles were measured as well in the case of the two high strength steels S690 and S960 (Figure 107 - Figure 110). For both longitudinal and transverse profiles, compressive stresses are met near the surface. Nevertheless, the transition point from the compressive to the tensile zone is shifted in this case closer to the surface and to a depth of 1 mm for both high strength steels and for both longitudinal and transverse RS. This is attributed to the shallower HFMI groove introduced in the case of the two high strength steels in comparison to S355.

It is becoming evident from Figure 107 to Figure 110 that analyses with non-viscous material modelling for both high strength steels exhibit very good agreement with the measured profiles, while those considering viscosity tend to overestimate the RS contrariwise to the case of S355. The non-viscous models of S690 predict overall the transverse RS with a deviation of 70 MPa or smaller. Only near the surface, a significant deviation of 200 MPa is met. In regard to the longitudinal profiles, very good agreement is met near the surface and up to a depth of 1.5 mm. Overestimation of the RS is evident for a depth between 1.25 mm and 1.75 mm. In the case of S960 similar or better agreement is met, especially for the case of longitudinal RS. On the contrary, the models considering strain rate dependency exhibit overall a significant deviation with the measured profiles of even higher than 100 % of the yield stress.

This incompatibility contradicts the above statement regarding the predominance of strain rate dependency during HFMI simulation. Nonetheless, tensile only material data was used for the calibration of the present model. The earlier-described influence of the deformation mode on the viscosity of steel and not a negligibility of the strain rate sensitivity could be accounted for this incompatibility. It is possible that for the high strength steels, the deviation between the tensile and the compressive strain rate dependency is much more significant, with the later one being significantly less viscous, almost strain rate independent. Therewith, the non-viscous model predicts RS much closer to the measured ones. In other words, when the static yield strength increases the strain rate dependency in compression is significantly reduced. This assumption is supported by the significantly better agreement of the models neglecting viscosity for both high strength steels and for both longitudinal and transverse RS. More experimental and numerical investigations in the future, which would confirm the above-stated assumption, are encouraged. It is as well possible that present material modelling is unsuitable for the current analyses and a more elaborate approach considering viscosity could lead to similar or better accuracy than that of the current analyses with non-viscous material behaviour.

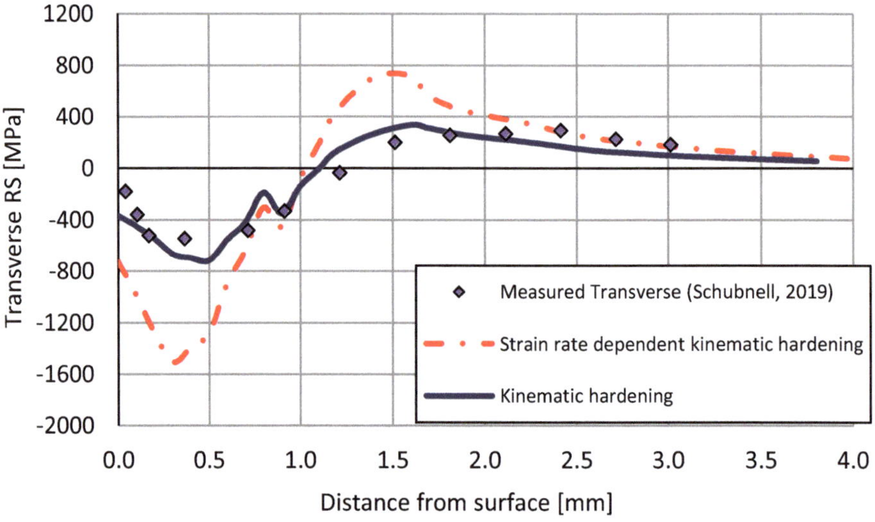

Figure 107: Simulated and measured transverse RS - Component of parent material S690 – (Schubnell, 2019) measurements were found in [145]

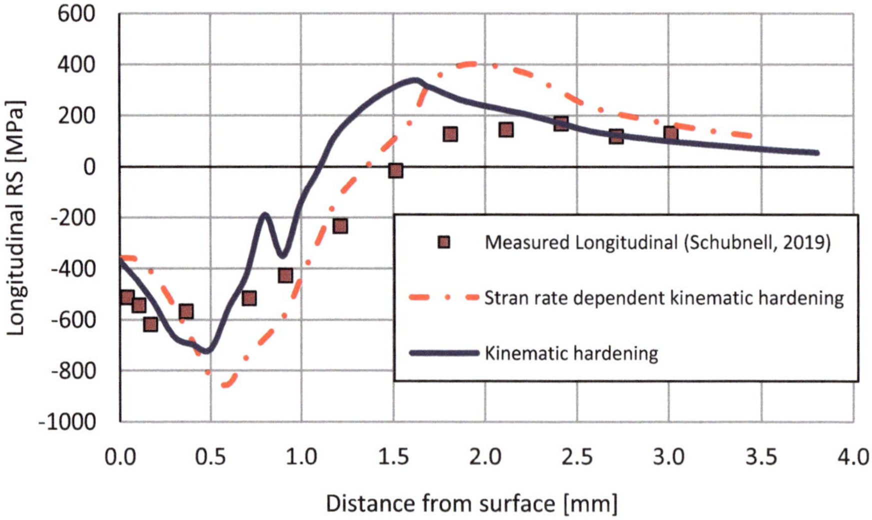

Figure 108: Simulated and measured longitudinal RS - Component of parent material S690 – (Schubnell, 2019) measurements were found in [145]

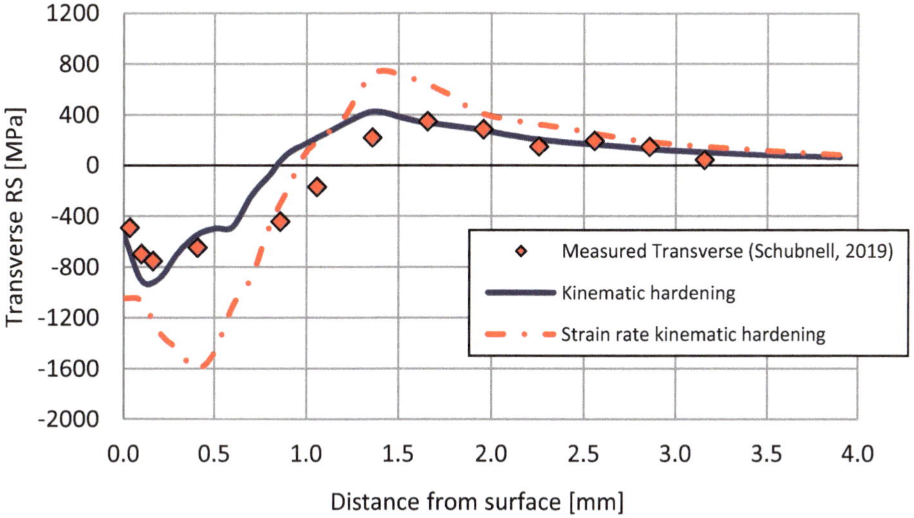

Figure 109: Simulated and measured transverse RS - component of parent material S960 – (Schubnell, 2019) measurements were found in [145]

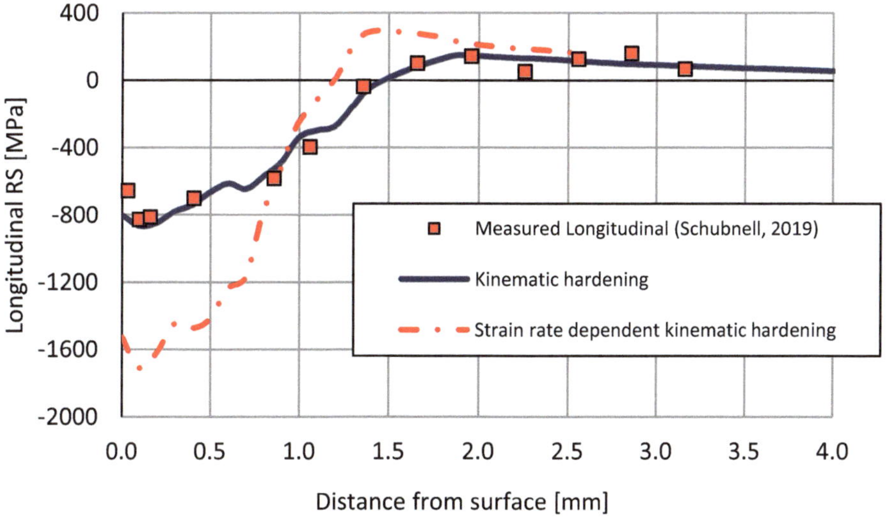

Figure 110: Simulated and measured longitudinal RS - Component of parent material S960 – (Schubnell, 2019) measurements were found in [145]

Additionally, the contour plots for the longitudinal and transverse RS of the S355 component are illustrated in Figure 111. They are presented exemplary so that a 3D overview of the stress field introduced upon, underneath and around the treatment surface can be enabled. The transverse contour plots, validate the theoretical ones with the zone of peak compressive RS to be found underneath the contact surface at a depth between 0.5 mm and 1 mm. In the case of the longitudinal stresses, the peak stress is met on the surface though[9]. Such a RS distribution was not described by theoretical profiles, as in most cases they are referring to RS fields of simulated or measured single impacts. Qualitatively similar plots were met in all the investigated cases, with only the magnitude and depth of the peak stresses differentiating.

With regard to the modelled treatment length of 10 mm, it is becoming evident in both cases that a homogenous stress field is created at the middle of the component. As it is expected, irregularities are met only at the beginning and end of the treatment line, where the initial and final impacts take place. This area of non-homogenous stress field is approximately equal to the diameter of the pin for the longitudinal RS and significantly smaller for the transverse ones. Moreover, significant RS arise away from the treatment area only for a very small distance, which in the case of the transverse RS is once again approximately equal to the diameter of the pin and for the longitudinal ones significantly smaller. Additionally, negligible RS are met near the bottom of the component, whereby the restraints are applied, or on the corners of the component, revealing that significant distance between them and the boundaries of the treatment line has been applied. Therefore, significant restraint or edge influence on the modelled RS is excluded.

The above remarks validate the initial assumption that the effect of the treatment is significantly localised. Therewith, it can be safely suggested that the simulation of a treatment length of three times the diameter of the pin or longer, will produce away from the treatment start and stop areas similar RS fields independently of the modelled treatment length. In other words, the treatment of a very long weld toe can be scaled down to a significantly smaller component, which takes into consideration the local geometry transverse to the treatment line. It can be as well rationally assumed that the restraints, which are applied to the real component under treatment away from the treatment area, have negligible effect on the introduced RS profile. It has to be highlighted that the simulation of the 10 mm treatment on the present 20 mm x 10 mm x 10 mm component,

[9] confusion by a direct comparison with the previous RS profiles should be avoided as both measurements and simulations were referring to averaged values in the depth direction, as it has been already mentioned

which would last in reality 2.5 s, lasted 7 hours 40 minutes 31 seconds using 256 processors with Massively Parallel Processing (MPP).

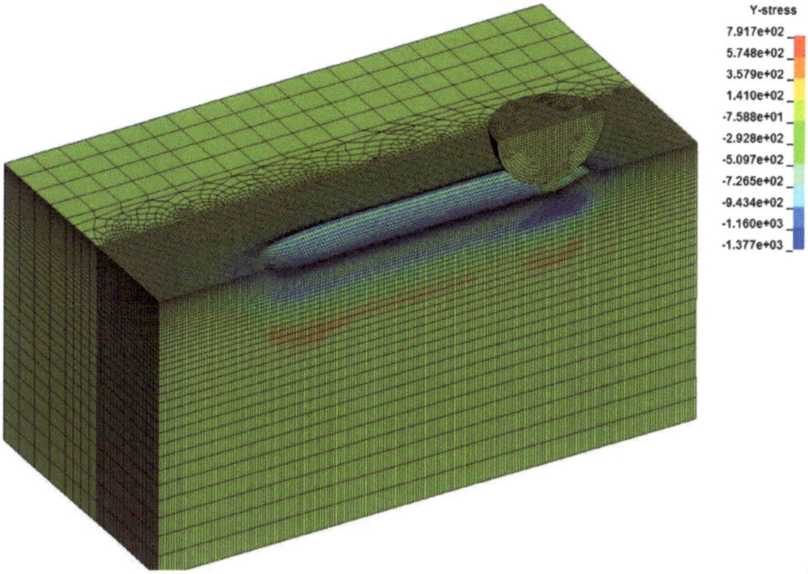

(a)

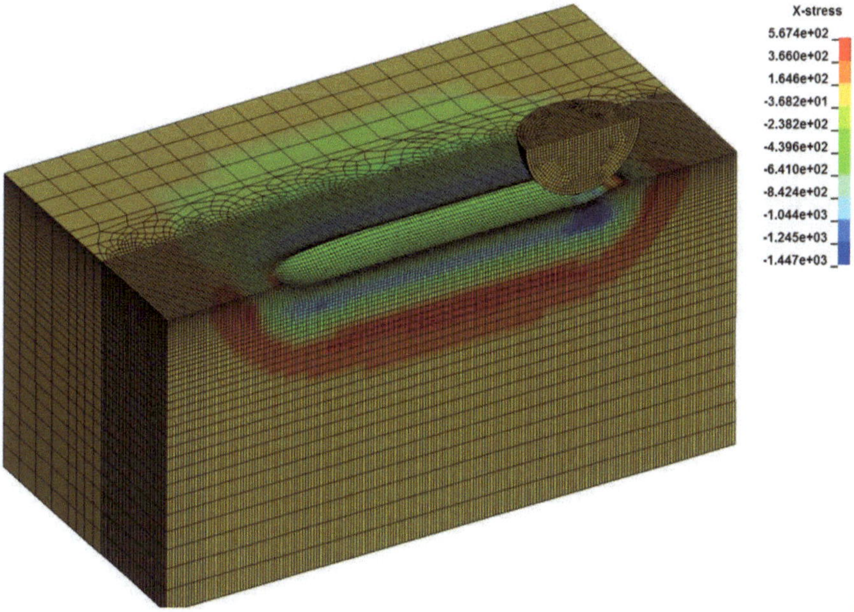

(b)

Figure 111: Stress contour plots of the specimen of parent material after the end of the treatment for the strain rate dependent case, calibrated based on compressive material testing, stresses are given in MPa: a) Longitudinal (Y); b) Transverse (X)

The significance of the above-stated conclusion regarding the scaling effect and the following assumption for the influence of the boundary conditions is of utmost significance, if it is considered that for real scale components treatment lengths of more than 1 m in total can be met[10]. As a full-scale modelling of such a component's treatment would not be possible, simulation of the treatment on the expected location of the first fatigue crack of the HFMI-treated component instead, could offer a sufficient estimation of the increased fatigue life, provided that the above assumptions are indeed valid. Further investigation on this scaling effect in future work is therefore encouraged[11].

5.3.1.5 Conclusions

The conclusions, which were drawn from the simulation of the specimens of parent material, are summarized as follows:

- During the simulation of HFMI, a mesh size of 1/40 of the pin's diameter or smaller is needed in order to provide mesh independent results for the current mesh type (linear brick solid elements).
- A displacement-based approach for the simulation of the pin's movement is proven unsuitable for high preciseness simulations of HFMI under the present conditions. Coupling of this approach with a strain rate dependent material model could provide better results, if it would be based on elaborate measurements and based on the assumption of a sinusoidal movement of the pin.
- A velocity-based approach for the simulation of the pin's movement is proven suitable for high preciseness simulations of HFMI. Measuring of the HFMI pin's contact force is straightforward and the calibration of the numerical model is not complicated. Moreover, such an approach is closer to reality than the displacement-based one.
- The present numerical results show satisfying agreement with the measured RS profiles. Compressive stresses are met near the surface constantly

[10] if for example, all the weld toes of a welded structural beam are hammered

[11] efforts for analogously scaling-down the welding simulation of large components by releasing the accuracy requirements have been made parallel to the present study in another research project of the KIT Steel and Lightweight Structures Institute [102]

reducing away from it until they pass to the tensile region in a depth of approximately 1.5 mm to 2 mm. In the case of the transverse residual stresses, a peak is met in the compressive zone below the surface at a depth of approx. 1 mm.

- The present simulated and measured RS profiles validate the theoretical ones provided for spherical indentation and those of previous analyses of HFMI and shot peening. The peak stress though, is met significantly deeper in comparison to the other methods, like shot peening. This shift is attributed to the repetitive treatment of the surface and the significantly deeper groove achieved by HFMI.
- According to initial expectations, the strain rate dependent material modelling enables a much more accurate simulation of the RS profiles for S355. On the contrary, the strain rate independent material models for the two high strength steels provided much better agreement, with the strain rate dependent ones overestimating the RS in all cases. It is possible that the significantly deviating strain rate dependence of these steels in tension and compression and the fact that the applied material model was calibrated based on tensile experimental data are causing this unexpected result. Further investigations are proposed to validate this assumption. Of course, it cannot be excluded that considering a more elaborate viscous material model could provide better agreement. Still, the agreement of the non-viscous models for the present first-step validation is satisfying in most cases.
- It has been shown that modelling a treatment length of more than 2.5 times the pin diameter, can create a homogenous stress field without singularities at distance larger than one pin diameter away from the treatment start and stop locations. Negligible influence from the external restraints was observed as well. The above observations support the assumption of the localized effect of the HFMI treatment and in extension the proposal that scaling down during a simulation of a long treatment can offer precise results for a larger component.

5.4 Fillet Welds

5.4.1 Investigated Component

Final step for the validation of the present modelling approach was the simulation of the fillet welds, which were investigated in the research project *HFH-Simulation* [146]. In the framework of this project, RS measurements were carried out for HFMI-treated components of the materials S355 and S960. Therefore, the same materials were investigated in this final step of the present study. The real components of the project were HFMI-treated at their initial, as-welded geometry (Figure 59, Figure 60). They were subsequently cut into smaller fatigue test specimens. Nonetheless, the simulation of the HFMI treatment at such scale was not possible due to computational restrictions. The treatment of a cut fatigue test specimen with length of 30 mm was modelled instead. Should the simulation provide results at the middle of the cut component, where the influence of stress relaxation due to cutting is negligible, comparable to the RS measurements[12], then the above-stated assumption regarding the scaling down of the component would be as well validated. The specimens, on which the RS profile were measured, were cut out from the middle of the parent fillet welds. Therefore, the modelled WRS from the middle of the component were introduced in the subsequent HFMI simulations. The investigated geometry for S355 is presented in Figure 112. In the case of S960, the symmetric component, which is presented in Figure 113, was taken into consideration.

5.4.2 Discretization

Discretization of the investigated component with a universal mesh size was not possible, as the available computational capacity would be exceeded. Therefore, two volumes with different mesh size were modelled for each specimen as it can be seen in Figure 112 and Figure 113. A volume, underneath the treatment line, was discretized with finer mesh according to the mesh requirements, which were suggested based on the investigation of parent material specimens. The rest of the component was discretized with coarser mesh. Compatibility equations of a

[12] especially for the transverse RS, which are being influenced significantly less by the cutting

bonded surface-to-surface connection [113] were modelled on the contact surfaces between the two volumes. A new and final convergence study was carried out in order to exclude any influence of this new discretization approach on the calculated RS.

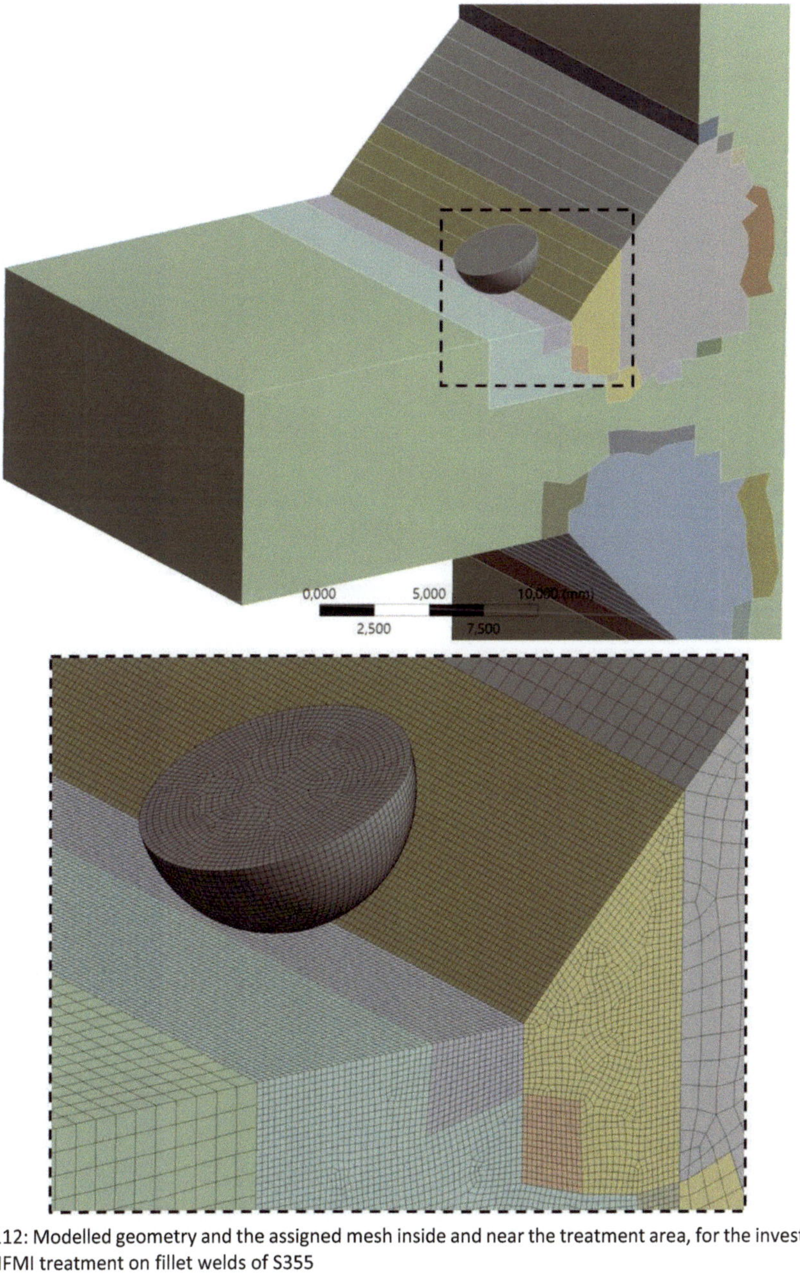

Figure 112: Modelled geometry and the assigned mesh inside and near the treatment area, for the investigation of the HFMI treatment on fillet welds of S355

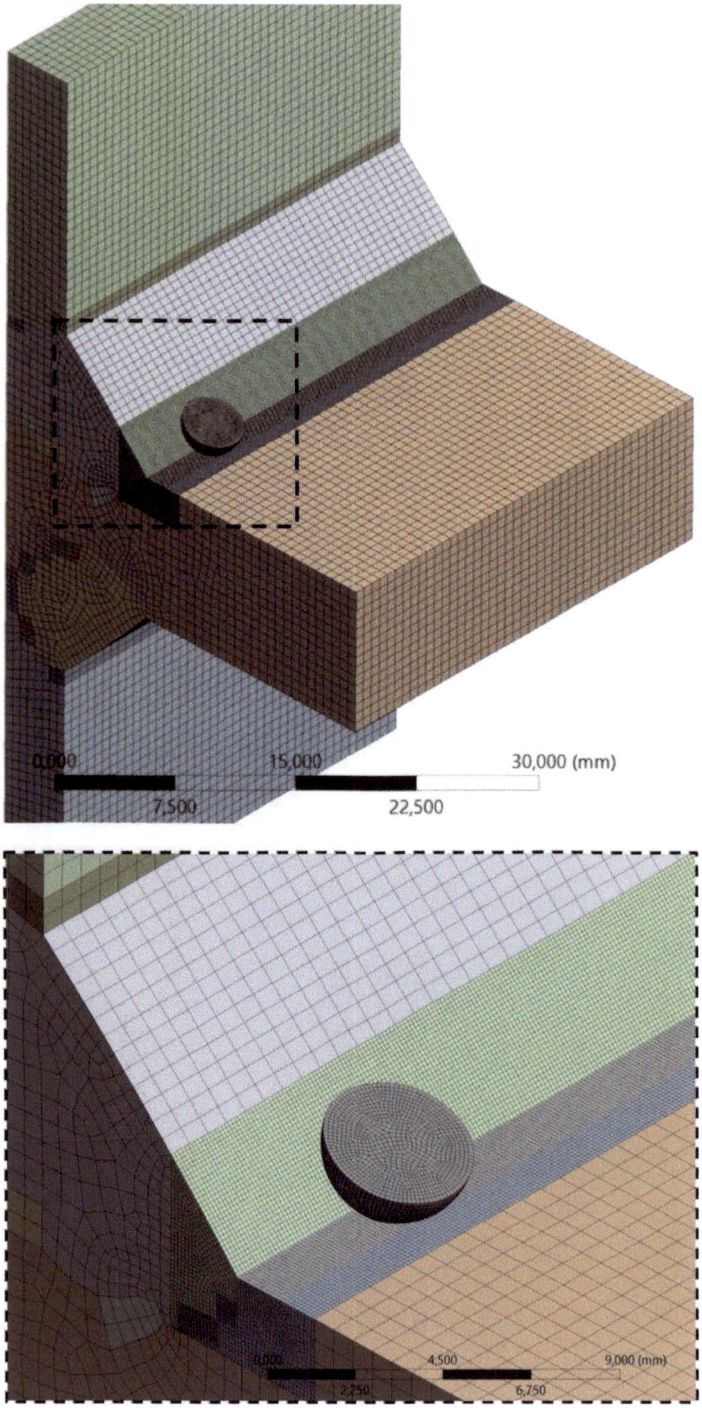

Figure 113: Modelled geometry and assigned mesh for the investigation of the HFMI treatment on fillet welds of S960

5.4.3 Modelling of Material Behaviour

Bilinear material behaviour with kinematic hardening and the Cowper-Symonds model (Eq. (12)) were once again taken into consideration for simulating respectively the elastic-plastic behaviour and the strain rate dependency of the investigated steels. Furthermore and in order to enable a more precise simulation, the differentiated microstructure of the HAZ and the FZ and their respective yield stress, as they were estimated during welding simulation, were introduced in the HFMI simulation model. The yield stress of the FZ was predicted earlier during the welding simulation and it was validated through the hardness measurements in [146]. PM yield stress was selected in all cases based on the above-presented respective material parameters. The yield stress of the HAZ and its strain rate dependency were explicitly calibrated through tensile tests under various strain rates of micro-specimens cut out from the HAZ [146]. The results of this material characterisation of the HAZ are presented in Table 44. The good agreement between the experimentally determined yield stress of the HAZ [146] and the respective values predicted through the present FE welding simulation, which are as well presented in Table 44[13], further supports the validity of the welding simulation approach, as it was presented in chapter 4. Moreover, the fact that satisfying agreement is met for all three investigated materials reveals once again the robustness and versatility of the approach. The calibrated parameters of the Cowper-Symonds model (Eq. (12)) for the HAZ were adopted for the rest of the material zones for all three materials with negligible expected influence on the results, as no significant strain rate is met in the PM and the microstructure in the FZ deviates slightly than that in the HAZ. *H* is assumed to be for all material zones equal to that of the parent material. The applied material parameters for the bilinear elastic-plastic behaviour and the Cowper-Symonds model and for all investigated materials are presented in Table 45.

[13] although as mentioned above, only S355 and S960 were simulated, the material parameters for S690 are presented as well

Table 44: Experimental results from [133] for the dynamic yield stress of the HAZ for various strain rates and predicted values of the static yield stress with the FE weld modelling approach

$\dot{\varepsilon}_p$ [s^{-1}]	σ_Y' [MPa] S355	S690	S690
FE (static)	786	861	981
0.001	713	861	952
1	856	961	1031
50	913	1003	1247
50	943	996	1198
500	1165	1126	1518
500	1200	1226	1457

5.4.4 HFMI Treatment Setup and Boundary Conditions

Setup of the HFMI treatment in the present simulation was based on the results of the investigations for the specimens of parent material, the recommendations of the manufacturers and the monitoring of the treatment on real specimens in [146]. The vertical axis of the pin was set at a 20° and 90° angle to the global vertical and longitudinal[14] axis. The centreline normal to the pin's upper circular face and coming through its centre was intersecting the weld toe at all times. Treatment was initially set to begin and end 5 mm away from the edges of the component. A local coordinate system was defined on the top circular face of the pin for implementing the respective boundary conditions. Initial position of the pin, its local coordinate system and the global one are presented in Figure 114.

[14] axis parallel to the treatment line

Table 45: Calibration of the Cowper-Symonds model based on the Data from Table 44

Material	Material Zone	σ_y [MPa]	*H* [GPa]	*Δ* [-]	*γ* [-]
S355	PM	450	0.9	5993	5
	HAZ	713			
	FZ	918			
S690	PM	734	0.9	11740	5
	HAZ	861			
	FZ	918			
S960	PM	1050	1.0	12830	5
	HAZ	952			
	FZ	963			

A velocity-based approach, as it was presented earlier for parent material specimens, was applied for simulating the movement of the pin vertical to its axis. The accelerating force is applied this time on the Z direction of the local coordinate system. The impact velocity, as it was determined for each material during the earlier investigations of PM, was initially adopted. Nonetheless, a change in the introduced impact velocity or a denser overlap of consecutive impacts was expected due to the altered hardness in the HAZ or the different local geometry at the weld toe, so that the real groove depth could be simulated. The trace depth was compared with real specimens for a second-step validation in a new *trial and error* calibration round. Finally, simulation of the treatment on all four weld toes was not possible with the present computational capacity. It was then decided to model the HFMI treatment only on the weld toe, where the RS were measured in [146]. Otherwise, when the simulation would be carried out predictively the treatment on the most unfavourable weld toe in terms of fatigue strength should be considered. As an ideal geometry is modelled in purpose, so that the influence of local weld geometric irregularities and other flaws is neglected, this would be the one with the highest tensile WRS.

Two pin diameters of 3 mm and 4 mm were considered in order to simulate the different RS profiles introduced by *HiFIT* and *PITec* [146]. Apart from the different pin diameter, the above-mentioned variations in frequency and contact force were as well taken into consideration. *PITec* treatment with a real average travelling step of 0.08 mm between consecutive impacts, was simulated in a 3-pass analysis, based on the results of the specimens of PM and previous studies ([44],

[45], [54] etc.). The measured average step of 0.22 mm of the *HiFIT* treatment between consecutive impacts dictated the simulation of a single-pass treatment.

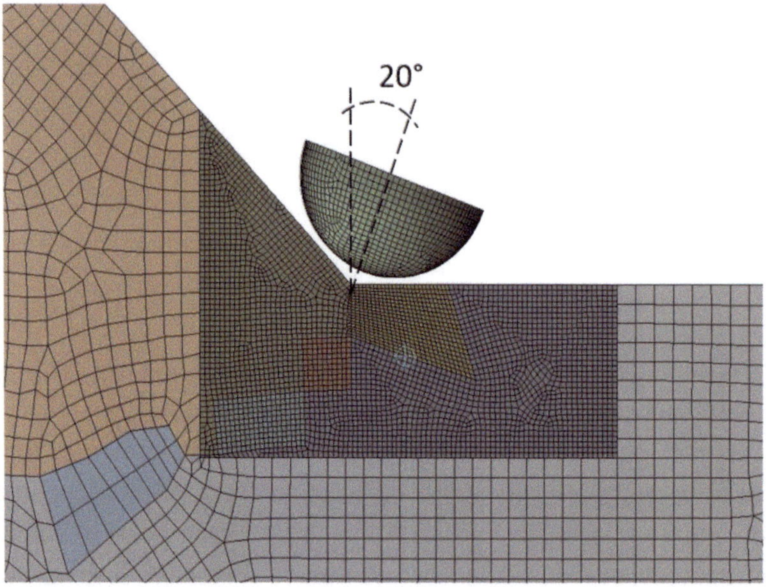

(a)

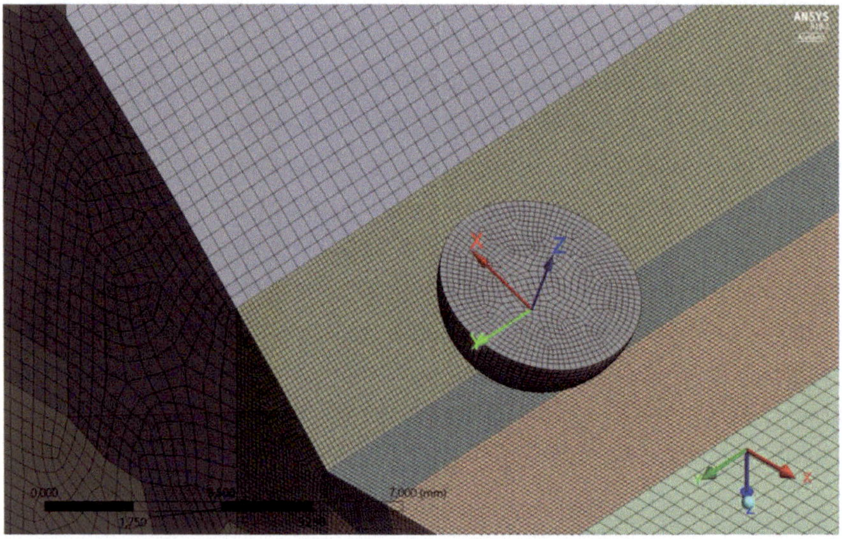

(b)

Figure 114: Initial position of the pin: a) Lateral view; b) Isometric view and the local and global coordinate systems

According to initial expectations, the introduction of the real groove depth with a single treatment pass would be attributed to the higher impact velocity of the

HiFIT device and the smaller pin diameter. Therewith, the difference between the two devices was taken into consideration during the present study.

5.4.5 Preliminary Investigations for the Simulation of Fillet Welds

Although mesh size requirements regarding the contact surface were already discussed, further and more complex effects could arise due to the introduction of two volumes with different mesh connected with contact surfaces and the more complicated geometry in the case of the fillet welds. Both of these changes could have had a significant effect on the propagation of the stress waves inside the component and in extension on the final RS field. Therefore, a new round of validation was carried out for this new mesh setup, as it was mentioned above. The symmetric model of S960 (Figure 113) was selected for this series of preliminary investigations. Six models with different mean element size in the fine and the coarse mesh volumes and variating distance between the boundary of the HFMI groove and the contact surface in the PM were compared. In all cases, a relatively high impact velocity of 4.8 m/s was applied so that excessive deformation would be introduced and the results would lie on the safe side. This velocity was same for all models so that the introduced groove would always have the same width and the influence of the distance between the groove boundary and the contact surface could be isolated. Three different values of 2.8 mm, 4.2 mm and 5.1 mm, for this distance measured from the weld toe, were considered. Two different element sizes of 0.125 mm and 0.1 mm and of 7 mm and 3.5 mm for the fine and the coarse mesh volumes respectively were introduced. Strain rate dependency was neglected and the material parameters for elastic-plastic behaviour of the material zones of the S960 weldment from Table 45 were adopted. An overview of the investigated models is presented in Table 46. The convergence of the transverse RS profile at the middle of the specimen perpendicular to the weld toe was adopted as the criterion for the selection of the appropriate mesh setup.

Table 46: Convergence analysis for the simulation of fillet welds

Model	Distance of the mesh transition surface from the crater's rim [mm]	Mean element size in the fine mesh area [mm]	Mean element size outside the fine mesh area [mm]
MS1	2.8	0.125	0.7
MS2	4.2	0.125	0.7
MS3	5.1	0.125	0.7
MS4	5.1	0.1	0.7
MS5	4.2	0.1	0.35
MS6	5.1	0.1	0.35

The results of this final convergence study are presented in Figure 115. *MS1* produced a profile that validated the theoretical transverse RS profiles. Nonetheless, an obvious discontinuity at the area around the transition area was met. When the contact surface was shifted further away from the crater's rim with *MS2* and *MS3* this influence became more and completely negligible respectively. Additionally, the RS profile inside and near the treatment, area changed significantly, with the two meshes *MS2* and *MS3* though, providing almost identical RS profiles with each other. Therefore, both distances of 4.2 mm and 5.1 mm from the weld toe for positioning of the mesh transition area were considered sufficient. A further refinement of the element size in the fine mesh area of the model *MS4* down to 0.1 mm though caused once again a significant change of the transverse RS profile. Moreover, a very significant stress concentration at the contact surface arose. As it was secured earlier that the distance between this surface and the crater rim is sufficiently large, the significant difference of element size on the two sides of this surface was accounted for this stress concentration. This dictated a further reduction of the element size in the coarse mesh. That was completed in the two final models of this convergence study, *MS5* and *MS6*, which provided almost identical results in the treatment zone and negligible stress concentration near the transition surface. *MS5* and *MS6* mesh setups consisted respectively of 1,236,739 and 1,188,710 nodes, requiring similar CPU time for their complete solution. Hence, the mesh setup MS6 was selected for all subsequent simulations. This distance of 5.1 mm between the treatment and transition area, had to be held to both transverse and depth directions though. As this would lead to a significant increase of the total number of the elements, adaptive mesh size inside the fine mesh volume was applied. Exemplary the final mesh for the simulation of the FW960 fillet weld is illustrated in Figure 116.

The above, proposed models were still marginally possible to be solved under the provided computational capacity. Nevertheless, increase of computational time due to the addition of strain rate dependency, further local mesh refinement due to later observations or consideration of WRS would exceed these limits. Ergo, it was decided to reduce the simulated component length from 30 mm down to 20 mm. The HFMI pin would start and stop its movement 5 mm away from the respective edges leading to a treatment length of 10 mm. Despite this reduction, the previously mentioned perquisites for scaling down, i.e. a minimum treatment length of 2.5 times the pin diameter and a distance of the treatment from edges of the components of 1 pin diameter, were met. With this length reduction and the above-mentioned finalized discretization setup the simulated model came up to a number of 1,255,205 elements for the case of FW960.

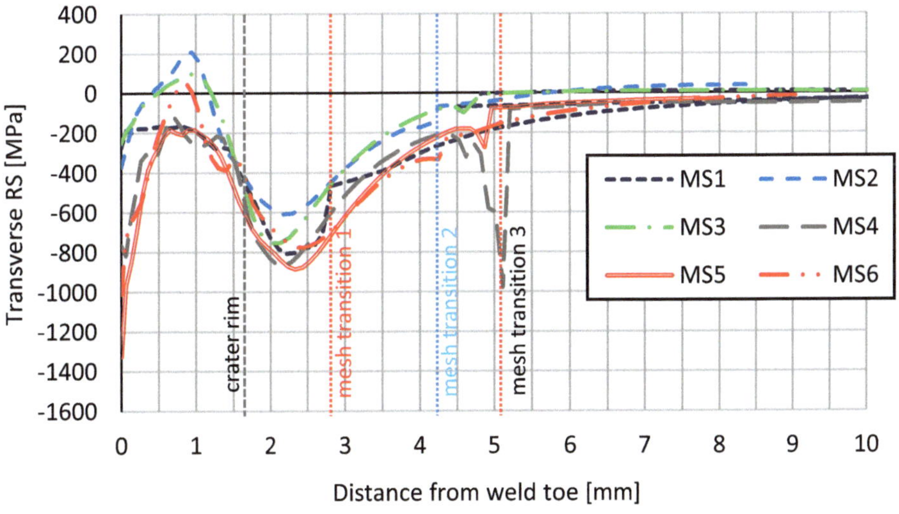

Figure 115: Results of the convergence study for the simulation of fillet welds

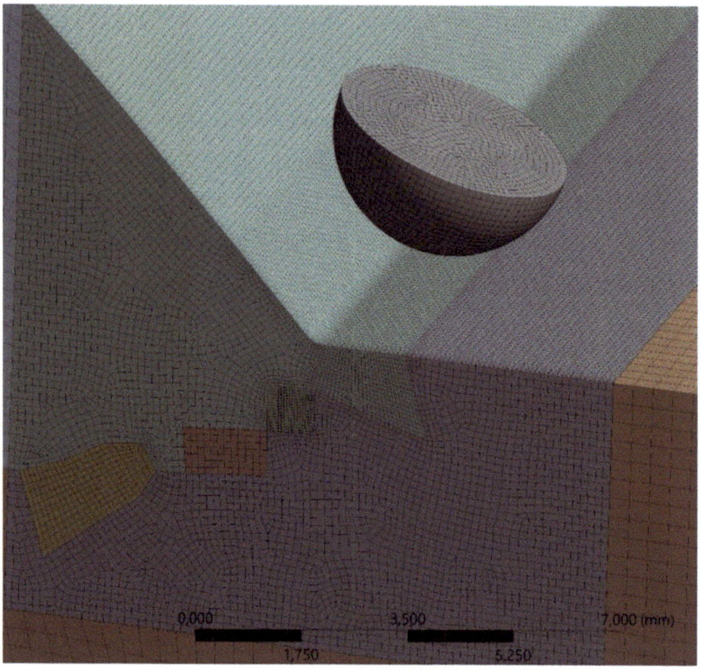

Figure 116: Final mesh for the simulation of FW960

5.4.6 Analyses and Results

5.4.6.1 Simulation of the HFMI Groove

Local Geometry at the Weld Toe

During preliminary investigations of the fillet welds, which were described above, it was observed that under the initially modelled geometry of Figure 117a, geometry and stress state of the weld toe after the treatment were unrealistic. An inner groove of a diameter of smaller than 0.1 mm was introduced at the bottom of the trace on the theoretical line of the weld toe. As it can be seen in Figure 22b, no such notch is met in the geometry of a real HFMI groove. This erroneous deformation was attributed to the sharp geometry of the weld toe and led to the introduction of an illogical, non-expected discontinuity on the stress profile transverse to the treatment trace, as it is illustrated in Figure 118. The contact between the pin and the component initiates at the flat areas around the weld toe and as they are pressed downwards, the weld toe tends to uplift. Due to the modelled sharp notch, this effect is accounted for the introduction of this smaller groove.

Therefore, it was decided to alter locally the geometry by adding a small fillet radius of 0.1 mm and refine the mesh so that at least 3 elements are assigned along this fillet (Figure 117b)[15]. Such a local, significantly smaller than the HFMI groove, change of the geometry can rationally be expected to have a negligible influence on final RS. The implemented change led to a significant improvement but not a complete elimination of the described phenomena, as it can be seen in Figure 117b. Therefore, a further increase of the filleting radius up to 0.2 mm and 0.5 mm was tested sequentially (Figure 117b and Figure 117c). After the latter change, the effect was completely eliminated. Additionally, the introduction of the fillets lead to a constant "normalization" of the RS profiles inside the treatment trace with the calculated RS profiles gradually converge to each other, at least qualitatively (Figure 118). It has to be underlined that during the present calibration, a single-pass treatment with a pin of 3 mm diameter was modelled. Such an impact speed was assigned to the pin that it would just reach contact with the weld toe. Calibration with the above setup lies on the safe side, as an increase of treatment passes or of the pin's penetration would cause a deeper groove and would possibly eliminate the above effect, even for the initial unfavourable geometry. Ergo, a filleting radius of 0.5 mm was adopted for all subsequent simulations.

[15] the reader should recall the fictitious 1 mm radius added to the weld toe during fracture mechanics simulations with the notch stress concept, a state of the art method [155]

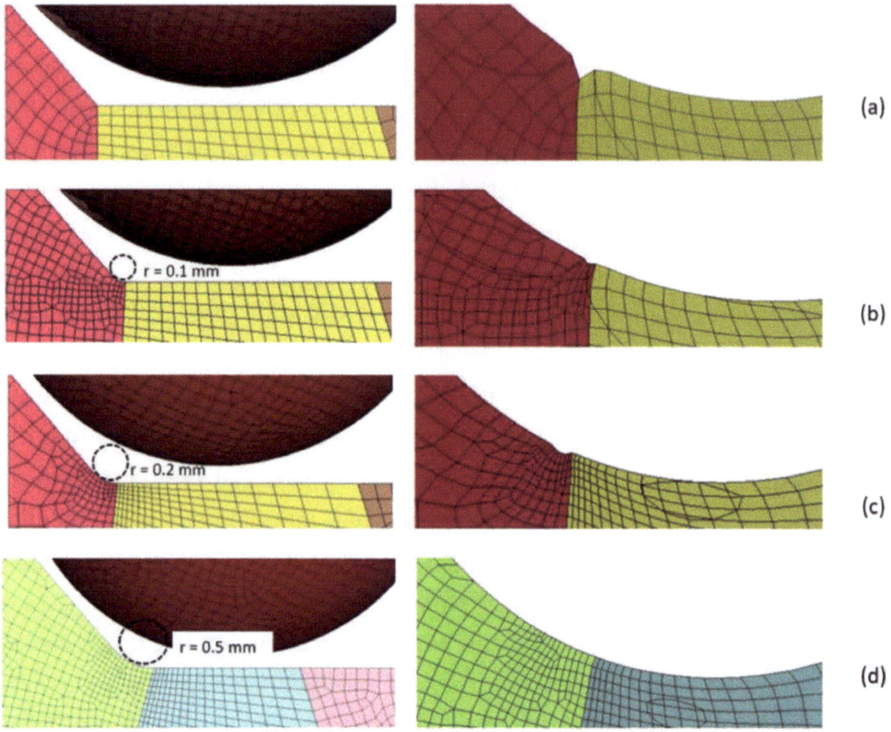

Figure 117: Initial geometry of the weld toe and the introduced HFMI groove after the 1st pass of the treatment: a) Without fillet; b) 0.1 mm fillet; c) 0.2 mm fillet; d) 0.5 mm fillet

Overlap and Impact Energy

As it was mentioned earlier, the impact velocity and travelling step of the HFMI pin from the investigations for parent material were applied initially in the simulation of the fillet welds as well. Nevertheless, during preliminary analyses the boundary conditions from previous investigations were proven to lead to reduced penetration. Additionally to the increased hardness in the HAZ and the FZ, under the present geometry, contact initiates simultaneously at the two points where the pin's outer surface is tangent to the plate and the weld seam. Therefore, the contact surface during and shortly after the initiation of the contact is double as in the case of the parent material, where initiation begun around the bottom point of the pin. Due to the presence of friction, this leads to more consumption of impact energy and in extension to a shallower trace. Moreover, it was noticed that the simulation of shallower HFMI groove introduced RS of unreasonably high magnitude in the treatment area. Hence, a new calibration round regarding the impact velocity and overlap step should take place.

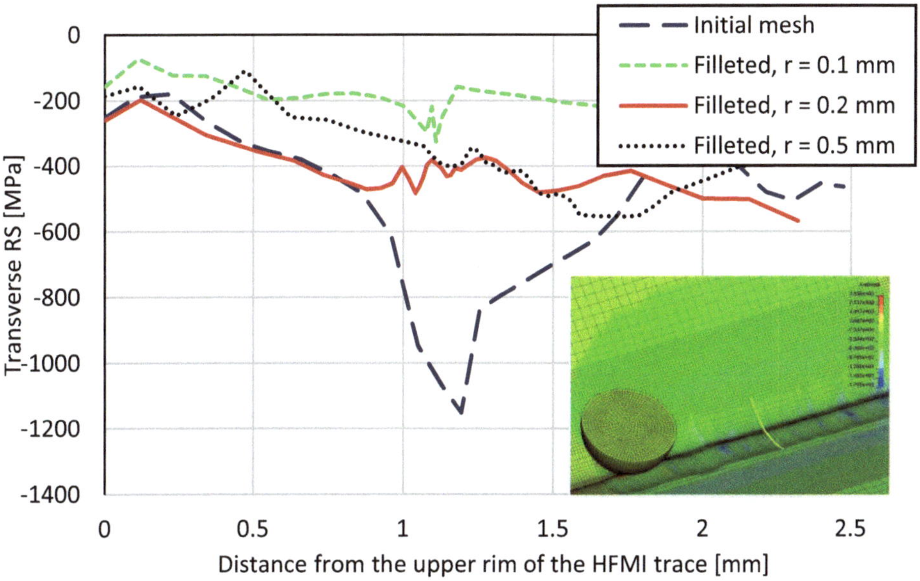

Figure 118: Transverse RS in the HFMI crater at the middle of the component perpendicular to the treatment

As the impact velocity applied during the simulation of the parent material was 3.27 m/s and already close to the mean measured impact velocity for the *PITec* device, it was decided to reduce the overlap between consecutive impacts. An increase of the velocity was implemented only in the cases, where a strain rate dependent material behaviour was considered and a further decrease in travelling step was no further numerically possible. Moreover, a higher impact velocity was selected for the cases of a 3 mm diameter pin, corresponding to the measurements on the *HiFIT* device. Several models were solved until the HFMI setup parameters for the accurate modelling of the HFMI groove could be selected. An overview is provided in Table 47. Based on the measurements from [146] and [168] and the proposal of IIW Recommendations [118] that the HFMI groove of 0.1 mm to 0.2 mm guarantees a complete treatment, introducing a trace with a maximum depth of 0.15 mm ± 0.025 mm was considered as a criterion for the selection of the appropriate boundary conditions for the specimens of S355. Moreover, it was considered that the vertical displacement at the weld toe should be at least 0.04 mm to exclude insufficient penetration, as at this point the vertical displacement exhibited a local minima in all cases. In the cases of S960 though, shallower grooves were measured overall. Hence, these criteria were adjusted to

0.12 mm ± 0.025 mm and 0.01 mm respectively. Exemplary, the vertical displacement inside the groove perpendicular to the treatment line for the case of a 2-pass treatment on a S355 specimen with viscous material behaviour and for an impact velocity of 4.2 m/s is given in Figure 119.

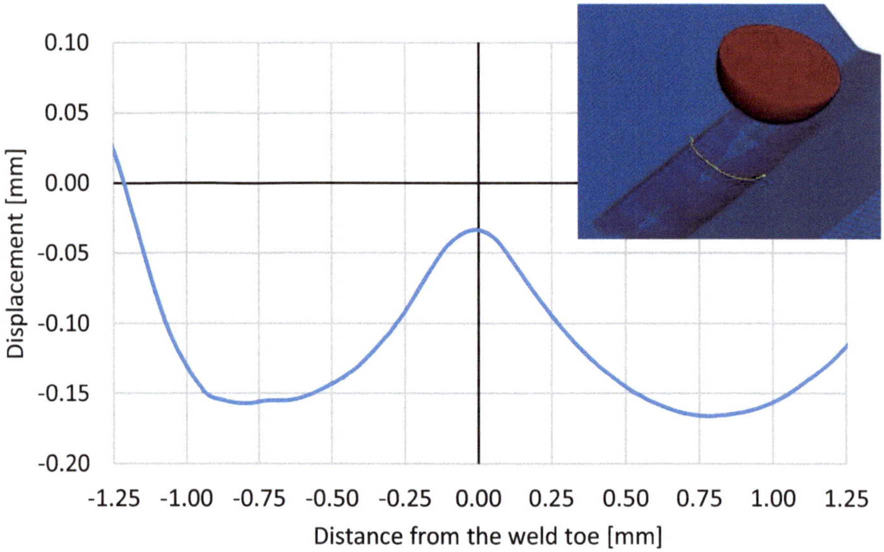

Figure 119: Displacement orthogonal to the upper plate surface inside the modelled HFMI groove perpendicular to the treatment line

It is becoming evident from the results of Table 47 regarding S355 that a travelling step of 0.2 mm between consecutive impacts of a single-pass treatment leads to insufficient penetration of the HFMI pin and the introduction of a shallow HFMI groove, even though a higher impact velocity of 4.2 m/s was considered. For a single-pass treatment with a 3 mm diameter pin, a reduction of the travelling step down to 0.15 mm and 0.1 mm was adopted for the non-viscous and viscous cases respectively, in order to achieve sufficient groove depth. Respectively, adopting a step of 0.15 mm in all models of a treatment with a 4 mm diameter pin led to sufficient results, both for non-viscous and viscous material behaviour. In the former case though, an impact velocity of 3.5 m/s, close to the respective one for the parent materials, was applied, while for the latter it was increased up to 4.2 m/s. Due to the increase of the impact velocity, only two passes were necessary for introducing a HFMI groove of sufficient depth. In the case of modelling the S960 fillet welds a reduction of the travelling step down to 0.1 mm was adopted in all cases, as once again a travelling step of 0.2 mm led to insufficient penetration. Impact velocity was in all cases equal to 4.2 m/s apart from the non-viscous model of the 3-pass treatment with a pin of 4 mm diameter. For this case, only two passes of treatment were necessary for the introduction of sufficient groove depth.

Table 47: Tested simulation setups for modelling of the HFMI trace

	Material behaviour	Pin diameter [mm]	Number of passes [-]	Step [mm]	Impact velocity [m/s]	Maximum groove depth [mm]	Vertical displacement at the weld toe [mm]
S355	non-viscous	3	1	0.2	4.2	0.10	0.01
	non-viscous	3	1	0.15	4.2	0.13	0.04
	viscous	3	1	0.15	4.2	0.09	0
	viscous	3	1	0.1	4.2	0.15	0.06
	non-viscous	4	3	0.15	4.2	0.29	0.15
	non-viscous	4	3	0.15	3.5	0.19	0.06
	viscous	4	2	0.15	4.2	0.17	0.05
	viscous	4	3	0.15	4.2	0.22	0.07
S960	non-viscous	3	1	0.2	4.2	0.12	0.01
	non-viscous	3	1	0.1	4.2	0.13	0.04
	viscous	3	1	0.1	4.2	0.09	0.01
	non-viscous	4	3	0.1	4.2	0.25	0.08
	non-viscous	4	2	0.1	3.5	0.22	0.06
	viscous	4	3	0.2	4.2	0.11	0
	viscous	4	3	0.1	3.5	0.13	0.01

It has to be underlined that for the present series of analyses, the impact velocity was modelled with a reduced preciseness of ± 0.25 m/s than in the case of the specimens of parent material. Due to the more complex local geometry at the weld toe and its influence on the elastic springback, calibration of the velocity-based model with a preciseness of ± 0.1 m/s was not practical any more. Moreover, due to the same effect few individual peaks of higher impact velocity, close to 5.5 m/s were met inside each treatment pass (approximately 3 peaks every

100 impacts), which could not be completely excluded with the current modelling approach. This led in some cases to the introduction of locally deeper HFMI groove and higher magnitude of RS. These relatively small, in comparison to the total groove surface, areas were excluded from the averaging techniques and the overall evaluation of the presented FE analyses. As their effect was significantly local, they did not influence the global RS field. The final simulations of the RS fields introduced by the HFMI treatment on the investigated fillet welds were based on the above remarks.

5.4.6.2 Mapping of Welding Residual Stresses

Mapping of WRS and the respective plastic strains was carried out in *ANSYS Workbench*, based on a *Profile Preserving* approach, with a *Triangulation Weighting* and a *Volumetric Transfer Type* [4]. A suitable algorithm was written in *MATLAB* [121] in order to enable the transformation of the output from the mapping process into *LS-DYNA* input code. This was necessary as due to the new fine mesh of the HFMI analysis large data files exceeding 10,000,000 command lines have been developed.

Figure 120: Contour plot of transverse WRS at the beginning of the HFMI simulation – Stresses are given in MPa

The transverse WRS contour plot at the initiation of the HFMI simulation in *LS-DYNA* is illustrated in Figure 120. The respective profile at the middle of the HFMI-treated component during the simulation initiation and after the first and second impacts are presented in Figure 121. After mapping of the WRS on the HFMI model, a slight, negligible stress relaxation was observed in most cases. This was initially expected though, as an even smaller component than the cut specimen of chapter 3.4.4.2 is simulated in this case. Nevertheless, the relaxation is considered negligible. Moreover, by comparing the three RS profiles of Figure 121, it is becoming evident that a redistribution of RS takes place even after the first HFMI impacts due to the propagation of elastic stress waves inside the component and the new equilibrium that arises after their superposition with the existing RS. Although the first and second impacts take place only at +0.2 mm and +0.4 mm after its initial position and 4.8 mm and 4.6 mm away from the line, where the RS of Figure 121 are evaluated their influence is becoming already evident.

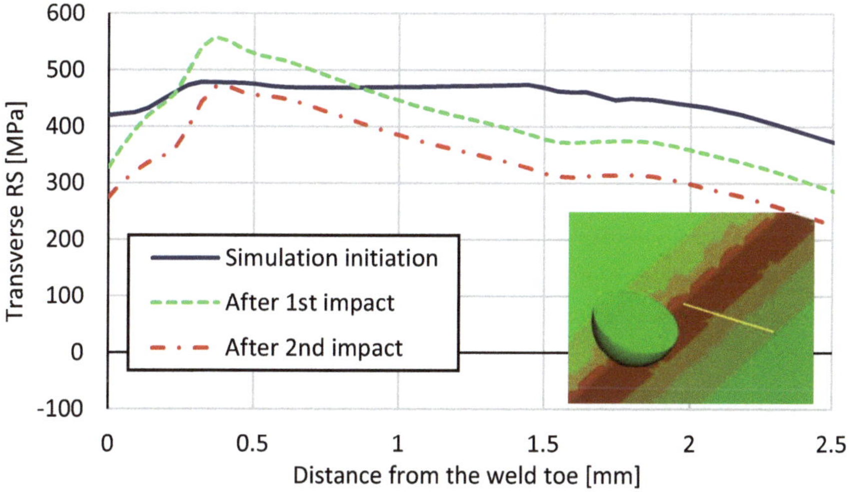

Figure 121: Transverse RS profiles at the middle of the component on the side of the HFMI simulated weld toe

5.4.6.3 Simulated RS profiles

A final series of RS simulations were carried out for the investigated fillet welds. As it was mentioned earlier, main goal was the validation of the present approach, for modelling RS from HFMI. In order to investigate the influence of neglecting or considering the WRS and the strain-rate dependency and the significance of simulating a realistic HFMI groove depth, more than one analyses were carried out for each material. The respective material parameters of Table 45 were applied in each case. The simulated RS profiles were compared with measurements, which were carried on the real fillet welds in the framework of [146]. Unlike the WRS measurements, which were carried out on the parent, uncut fillet welds, the present measurements were performed on the cut specimens of [146].

Exemplary one specimen treated with a *HiFIT* and one with a *PITec* device were measured for each material.

The measurements were carried on the surface and in depth direction with an X-Ray diffractometer of the IWM and with neutron diffraction in the *Helmholtz Zentrum* (HZB), Berlin, Germany respectively. Only transverse RS were measured in the former one, while both transverse and longitudinal were investigated in the latter. Two measurements were carried out on the surface at the middle of the specimen and perpendicular to the HFMI treatment line, beginning from the centre of the HFMI groove and with a step of 0.5 mm. The collimator diameter was 2.0 mm. In the case of the neutron diffraction, the measurements were once again carried out at the middle of the component and underneath the weld toe, which coincided with the middle of the HFMI groove. Gauge volumes of 2 mm x 2 mm x 2 mm and 5 mm x 2 mm x 2 mm were investigated for the longitudinal and transverse direction respectively. Measurements were carried out every 0.5 mm in depth direction. The applied gauge volumes were significantly larger than those, which were measured earlier in [44] and [145] for the specimens of parent material. Once again, averaging of the results from the integration points on the measured surface or inside the measured gauge volumes was necessary in order to enable a direct comparison of simulation and measurements. More details regarding the present RS measurements can be found in [144] and [146].

S355 Specimens Treated with a Pin of 3 mm Diameter

An overview of the numerical investigations regarding the single-pass treatments of the S355 fillet welds with a 3 mm diameter HFMI pin is given in Table 48. Three analyses, *A1, A2* and *A3*, adopting a non-viscous material behaviour were carried out, with the first one neglecting and the other two considering the WRS. Due to a larger travelling step of 0.2 mm between consecutive impacts, *A2* introduced a HFMI groove shallower than the above-mentioned criterion of acceptance. Similarly, the model *A4* considering strain-rate dependency and a step of 0.15 mm led to the introduction of a shallow trace. An adoption of a smaller step of 0.1 mm in *A5* produced an acceptable HFMI groove. All simulations were carried out with an impact velocity of 4.2 m/s. The HFMI groove in the present case had an average half width of 0.8 mm.

The simulated transverse RS profiles on the surface of the fillet welds and perpendicular to the HFMI groove are presented along with the respective measurements in Figure 122. A non-negligible deviation is met between the two measurement series only at a distance of 3 mm from the weld toe. The agreement on the rest of the measurement points is very good. They both produce qualitatively similar results; compressive RS are met along total measurement length and the peak stress is met outside the rim of the HFMI groove at a proximity of approximately 2 mm and 2.5 mm respectively. Peak stress of approximately -300 MPa is measured in both cases. Inside the HFMI groove though, significantly lower stresses are documented approaching zero at its middle.

Nonetheless, all simulated RS profiles, which are qualitatively similar to each other, show a peak stress at the middle of the HFMI groove. Models *A1, A2, A4* and *A5* significantly overestimate the RS inside the treatment area. They all approach the measured profile though, outside the HFMI groove. On the contrary, model *A3*, which predicted better the order of magnitude of the measured RS inside the HFMI groove, underestimates the RS away from the trace boundary. *A1, A2 and A4* are expected to show a significant deviation as they respectively neglect the WRS and introduce a HFMI trace shallower than the real.

Table 48: FE Analyses of the single-pass HFMI treatment on fillet welds of S355 with a pin diameter of 3 mm

Model	Material behaviour	WRS	Step [mm]	Impact velocity [m/s]	Maximum groove depth [mm]	Vertical displacement at the weld toe [mm]
A1	non-viscous	neglected	0.15	4.2	0.13	0.04
A2	non-viscous	considered	0.2	4.2	0.10	0.01
A3	non-viscous	considered	0.15	4.2	0.13	0.04
A4	viscous	considered	0.15	4.2	0.09	0
A5	viscous	considered	0.1	4.2	0.15	0.06

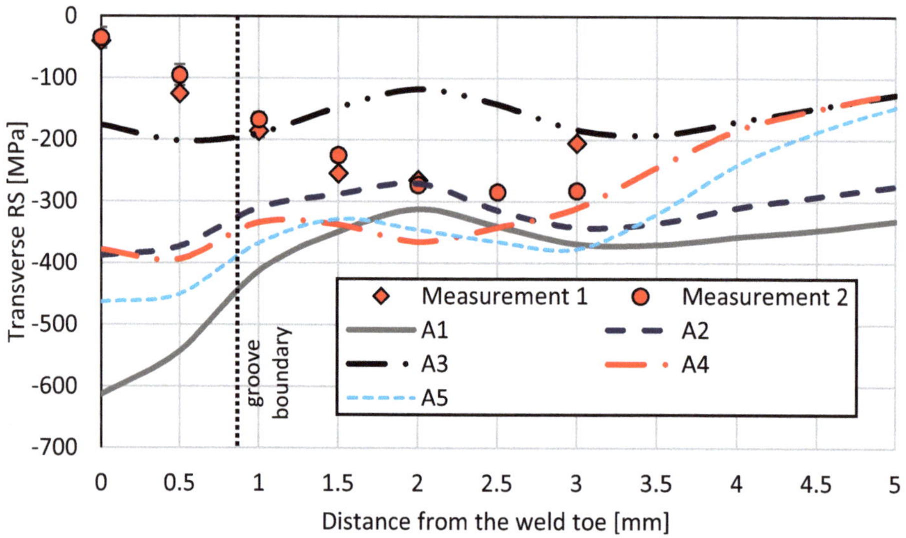

Figure 122: Simulated and measured transverse RS profiles on the surface of the S355 specimen treated with a pin of 3 mm diameter, perpendicular to the HFMI groove – Measurements from [146]

The simulated transverse and longitudinal RS profiles in depth direction are presented along with the respective measurements in Figure 123 and Figure 124. In the case of the transverse RS, analyses considering a non-viscous material model match qualitatively the measured profiles, being in the compressive regime near the surface and passing to the tensile one after a significant depth of approximately 3 mm or larger. The models considering strain-rate dependency produced different profiles even qualitatively. Once again, the introduced RS near the surface are compressive but the transition to the tensile zone takes place at much smaller depth of around 1 mm to 1.5 mm. The compressive stresses near the surface though have a higher magnitude, than those simulated with non-viscous material modelling. This change due to the introduction of strain rate dependency is expected as the elements on the top layers, underneath the contact surface, yield at a higher stress. Due to the higher magnitude of compressive residual stress and as the energy of the system is equal with the non-viscous case, the zone of the compressive stresses has to become thinner so that equilibrium of the RS can be achieved. Therewith, the transition points shifts up. Regarding the quantitative agreement of numerical and experimental results, model *A3*, which neglects strain-rate dependency but considered the WRS, shows overall a very good agreement with the measured RS. The highest deviation of 60 MPa, which is lower than

10 % of the material's yield stress, is met near the surface. Models *A1, A2* and *A4* are expected to show a significant deviation due to the above-stated reasons. The profile of *A5* though, despite considering the WRS and introducing a realistic HFMI groove, still deviates significantly from the measured one.

In the case of the longitudinal RS (Figure 124), all numerical models produce profiles qualitatively similar with each other and with the measured ones. In all cases, only compressive stresses are introduced in the first 4 mm underneath the contact surface. Model *A3* once again exhibits the best agreement with the measurements especially near the surface. A significant deviation of 100 MPa is met only at a depth between 1 mm and 1.5 mm. Otherwise the deviation is considered overall acceptable. The rest of the profiles although less accurate than *A3* deviate slightly from it, with each other and with the measured RS. This fact may reveal a smaller sensitivity of the longitudinal RS to the investigated effects in comparison to the transverse, especially regarding the depth of the introduced HFMI groove.

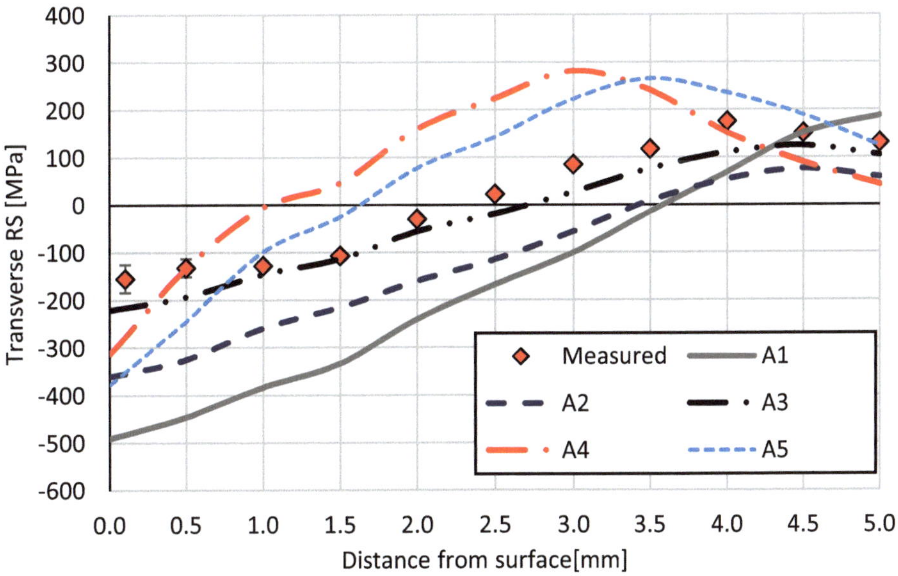

Figure 123: Simulated and measured transverse RS profiles in depth direction at the weld toe of the S355 specimen treated with a pin of 3 mm diameter, at the middle of the specimen – Measurements from [146]

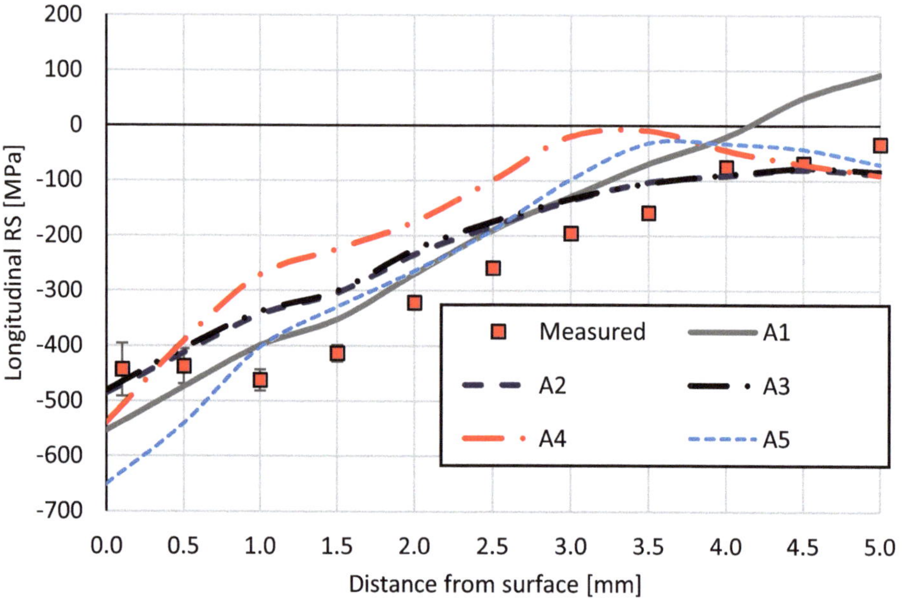

Figure 124: Simulated and measured longitudinal RS profiles in depth direction at the weld toe of the S355 specimen treated with a pin of 3 mm diameter, at the middle of the specimen – Measurements from [146]

Exemplarily, the applied mesh and the longitudinal and transverse RS contours of the model *A3* below the contact surface at a section at the middle of the investigated component and perpendicular to the introduced HFMI groove are presented in Figure 125. In the case of the transverse RS the compressive zone underneath and around the contact surface is becoming evident, without exhibiting though at any point significantly higher stresses than the rest. Overall, the compressive stresses do not exceed the yield stress of the material in the treatment area. On the contrary, longitudinal stresses significantly higher than the yield strength, even higher than 1400 MPa, are met on the top layer elements underneath the contact surface. Nevertheless, the RS in the rest of the compressive zone are as well approximately equal to the real yield stress of the material in the HAZ. Significantly higher stresses met exclusively at the top layer of elements could be as well the result of the numerical modelling of contact. Away from the treatment zone, significant tensile stress of even up to 500 MPa are met. This field of tensile stresses, which is attributed to the initial WRS field, is interrupted by the compressive stresses introduced with HFMI. The redistribution of RS through the HFMI deformation in the area of the weld toe is becoming evident.

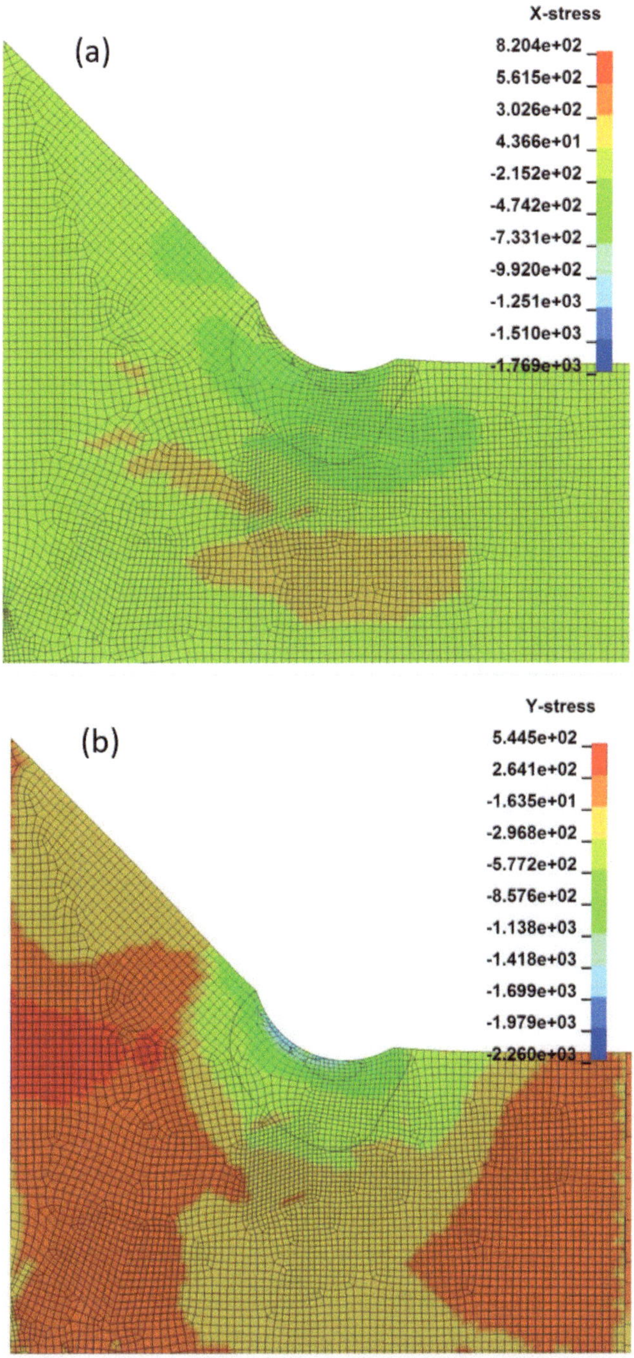

Figure 125: The applied mesh and RS contours of the model *A3* inside and near the HFMI groove – Section at the middle of the investigated component – Stresses are given in MPa: a) Transverse; b) Longitudinal

S355 Specimens Treated with a Pin of 4 mm Diameter

An overview of the numerical investigations regarding the treatments of the S355 fillet welds with a 4 mm diameter HFMI pin is given in Table 49. Four models were solved during the present investigation, two neglecting and two considering the viscosity of the material. Model *B1* neglected the WRS, while *B3* introduced a relatively deeper trace. *B2* and *B4* both considered the WRS and simulated the HFMI groove with better accuracy.

The simulated transverse RS profiles on the surface of the fillet welds and perpendicular to the HFMI groove are presented along with the respective measurements in Figure 126. Although a significant quantitative deviation is met between the two measurement series, they both produce qualitatively similar results. Compressive RS are met in both cases on all the measurement points and the peak stress is located at the boundaries of the HFMI groove at a proximity of approximately 1.5 mm from the weld toe. Peak stress of approximately -500 MPa is measured in both cases. Inside the HFMI groove though, a deviation of almost up to 300 MPa is documented.

Table 49: FE Analyses of the HFMI treatment on fillet welds of S355 with a pin diameter of 4 mm

Model	Material behaviour	WRS	Step [mm]	Impact velocity [m/s]	Number of passes [-]	Maximum groove depth [mm]	Vertical displacement at the weld toe [mm]
B1	non-viscous	neglected	0.15	4.2	3	0.29	0.15
B2	non-viscous	considered	0.15	3.5	3	0.19	0.06
B3	viscous	considered	0.15	4.2	3	0.22	0.07
B4	viscous	considered	0.15	4.2	2	0.17	0.05

On the contrary, all simulated RS profiles, which are qualitatively similar with each other, exhibit a peak stress at the middle of the HFMI groove. Models *B1, B3* and *B4* significantly overestimate the RS inside the treatment area. Nevertheless, as it can be seen from Table 47, the simulated treatment of *B3* introduces a much deeper HFMI groove than the real one and such a significant deviation of the RS profile is expected. The non-viscous material model *B1* predicts with very good preciseness the measured RS inside the HFMI groove as it coincides with the second measurement series. Nevertheless, as the distance from the weld toe is getting larger, the model underestimates significantly the measured RS with a deviation of even up to 250 MPa. The strain-rate dependent model *B4* on the contrary, lies sufficiently close to the measured RS away from the weld toe, despite the fact that the RS inside the HFMI trace are significantly overestimated.

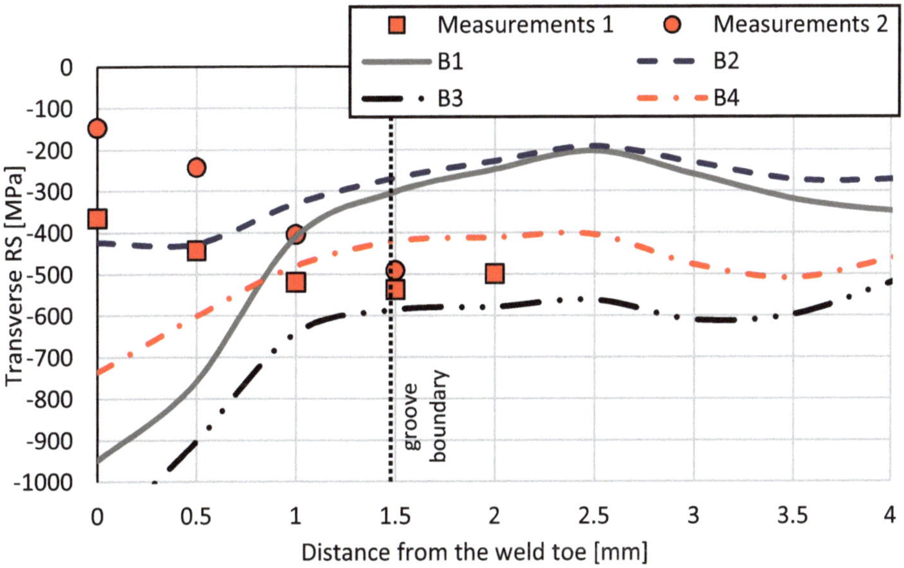

Figure 126: Simulated and measured transverse RS profiles on the surface of the S355 specimen treated with a pin of 4 mm diameter, perpendicular to the HFMI groove – Measurements from [146]

The simulated transverse and longitudinal RS profiles in depth direction are presented along with the respective measurements in Figure 127 and Figure 128. In both cases, all simulated RS profiles match qualitatively the measured ones, being in the compressive regime near the surface and passing to the tensile one after a significant depth of approximately 3 mm. Hence, significant quantitative deviations are found in most cases.

In the case of the transverse RS (Figure 127) the non-viscous models lie closer to the measured profile near the surface, while those considering strain rate dependency overestimated significantly the magnitude of the RS at this area. Nonetheless, from a depth of 2 mm and deeper the strain-rate dependent models *B3* and *B4* seem to simulate the measured RS profiles with high accuracy, modelling with preciseness even the transition point between the compressive and tensile zones. Still, the non-viscous material models lie much closer to the measured profile near the surface, overestimating the magnitude of the compressive RS at a larger depth. Therewith, the transition from the compressive to the tensile area is shifted deeper. The preciseness of the simulated transverse RS near the surface though, where crack initiation takes place, has to be considered as predominant of the appropriate modelling approach. From this point of view, the non-viscous model *B2*, which considered the WRS, exhibits the best agreement, with an overall deviation of up to 120 MPa or smaller. Still, the preciseness of this analysis is less in comparison to the above-presented case of single-pass treatment with a pin of 3 mm diameter simulated by model *A3*.

In the case of the longitudinal RS (Figure 128) the non-viscous models underestimate the compressive measured RS overall, with the largest deviation of almost 200 MPa met for *B2* near the surface. The viscous models and especially *B3* predict with sufficient preciseness the RS in the 1st mm near the surface, but as well underestimate the RS at larger depth.

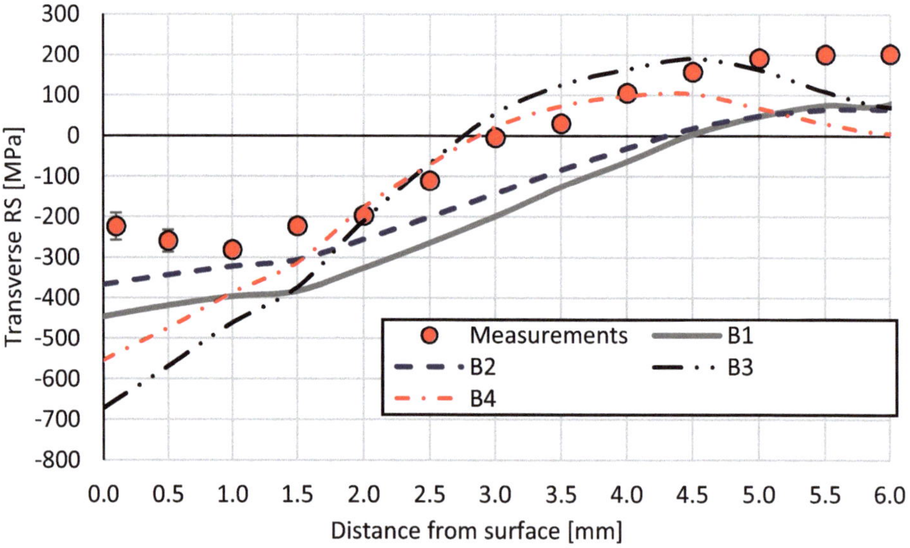

Figure 127: Simulated and measured transverse RS profiles in depth direction at the weld toe of the S355 specimen treated with a pin of 4 mm diameter, at the middle of the specimen – Measurements from [146]

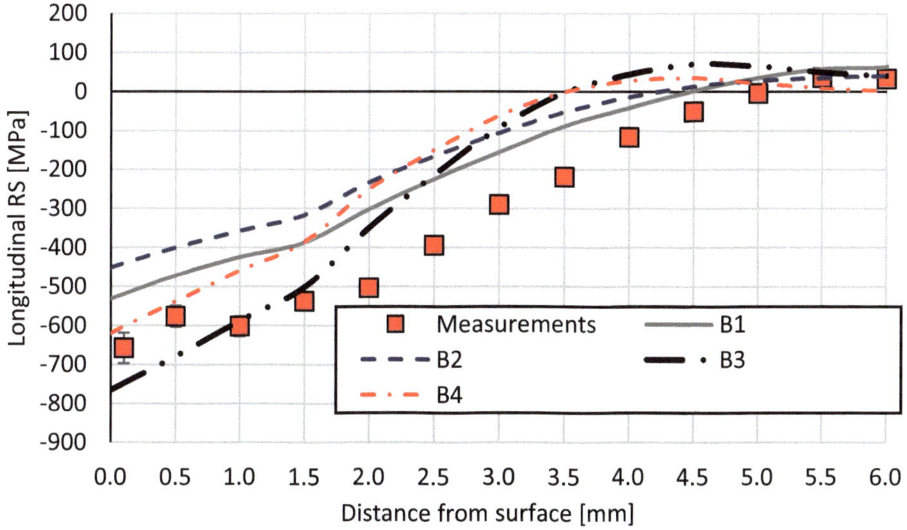

Figure 128: Simulated and measured longitudinal RS profiles in depth direction at the weld toe of the S355 specimen treated with a pin of 4 mm diameter, at the middle of the specimen – Measurements from [146]

Exemplary, the longitudinal and transverse RS contours of the model *B2* below the contact surface at a section at the middle of the investigated component and perpendicular to the introduced HFMI groove are presented in Figure 129. In the case of the transverse RS the compressive zone underneath and around the contact surface is becoming evident. According to initial expectations, the largest compressive RS are approximately 650 MPa and slightly lower than the local yield stress at the HAZ. They are met in an elliptical area underneath the contact surface. Overall, the compressive stresses do not exceed the yield stress of the material. Away from the treatment trace, lower tensile RS are met without significant peaks. On the contrary, high longitudinal stresses of up to 1400 MPa are met on the top layer elements underneath the contact surface once again, as in the case of a treatment with a pin of 3 mm diameter. Nevertheless, the RS in the rest of the compressive zone are as well approximately equal to the real yield stress of the material in the HAZ. The tensile stresses away from the treatment zone though, are lower in this case. The more significant redistribution of RS, which is met in this case, is attributed to the wider HFMI groove that is introduced by the wider HFMI pin and the higher number of treatment passes.

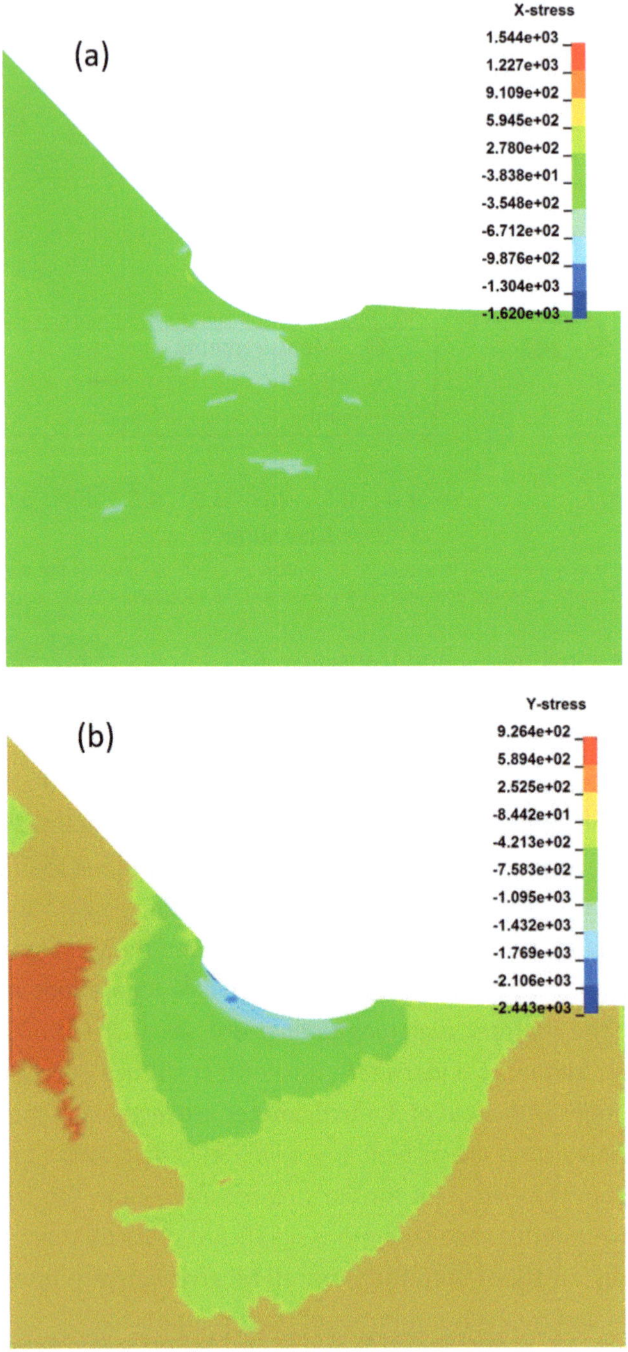

Figure 129: RS contours of the model *B2* inside and near the HFMI groove – Section at the middle of the investigated component, stresses are given in MPa: a) Transverse; b) Longitudinal

S960 Specimens Treated with a Pin of 3 mm Diameter

An overview of the numerical investigations regarding the single-pass treatments of the S960 fillet welds with a 3 mm diameter HFMI pin is given in Table 50. Four different models were considered in the present investigation, three neglecting and one considering the material's viscosity. Model *C2* due to the larger assumed step of 0.2 mm between consecutive impacts, introduced an unrealistically shallow HFMI groove, while *C1* neglected WRS. Therefore, these two models were expected to exhibit significant deviation from the measured RS. The HFMI groove in the present case had an average half width of 1 mm.

The simulated transverse RS profiles on the surface of the fillet welds and perpendicular to the HFMI groove are presented along with the respective measurements in Figure 130. A deviation of 100 MPa or less was met between the two measurement series for the whole measurement length, as they produce similar results both qualitatively and quantitatively. Compressive RS are met in both cases on all measurement points and the peak stress is met outside the boundary of the HFMI groove at a proximity of 2 mm and 3 mm from the weld toe. Peak stress of approximately -300 MPa is measured in both cases. Inside the HFMI groove though, lower stresses are documented lying close to zero at its middle.

Simulated RS profiles *C1* - *C3* produce results qualitatively similar to each other and with the measured profiles. They all exhibited lower stress inside the HFMI groove and a peak stress at approximately 2.5 mm away from the weld toe. Model *C3* approached the measured RS profile as well with high preciseness, showing a deviation of less than 100 MPa from the second measurement all along the measurement length. Models *C1* and *C2* as expected exhibited less accuracy than the model *C3*. Regarding the 4th carried out analysis *C4*, once again the introduction of strain-rate dependency in the considered material model led to a larger deviation between measurements and simulation results and even qualitative discrepancies.

Table 50: FE Analyses of the single-pass HFMI treatment on fillet welds of S960 with a pin diameter of 3 mm

Model	Material behaviour	WRS	Step [mm]	Impact Velocity [m/s]	Maximum groove depth [mm]	Vertical displacement at the weld toe [mm]
C1	non-viscous	neglected	0.1	4.2	0.13	0.04
C2	non-viscous	considered	0.2	4.2	0.12	0.01
C3	non-viscous	considered	0.1	4.2	0.13	0.04
C4	viscous	considered	0.1	4.2	0.09	0.01

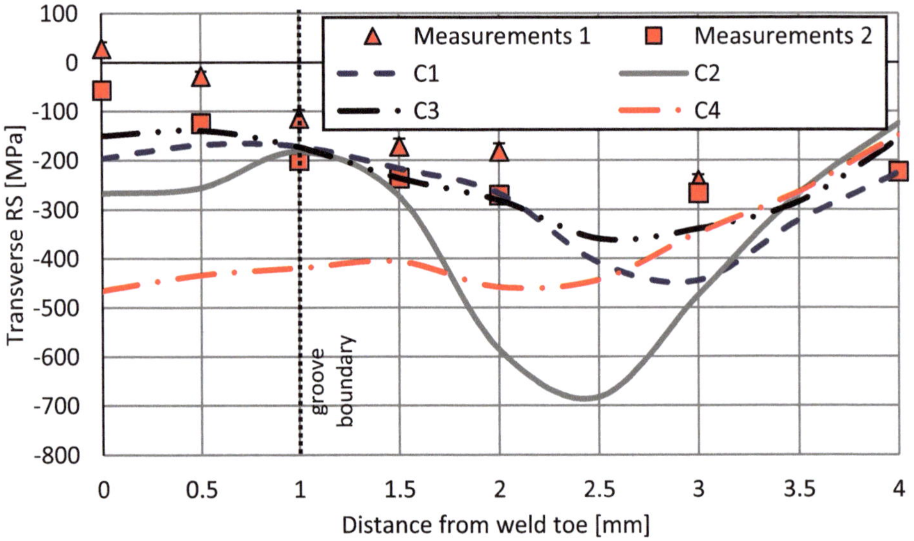

Figure 130: Simulated and measured transverse RS profiles on the surface of the S960 specimen, perpendicular to the HFMI groove – Measurements from [146]

The simulated transverse and longitudinal RS profiles in depth direction are presented along with the respective measurements in Figure 131 and Figure 132. In the case of the transverse RS, all numerical models produced profiles qualitatively similar with each other and with the measured one. According to the measurements, compressive RS are met in the first 1.2 mm underneath the surface changing to tensile for larger depths. Simulated profiles tend to shift from the compressive to the tensile regime at around 1.7 mm to 2 mm. Once again, analysis *C3* considering WRS but neglecting the strain-rate dependency estimated the measured RS with satisfying preciseness. Significant deviation of 110 MPa is met only at a depth of 1.5 mm or at depths larger than 3 mm. Still, the agreement of the model with the measured stresses near the surface is very good.

In the case of the longitudinal RS (Figure 131), all numerical models produce profiles qualitatively similar with each other and with the measured as well. In all cases, only compressive stresses are introduced in the first 3 mm underneath the contact surface. As in the previous analyses series *B1-B4*, the numerical models tend to overestimate the depth of the point, where transition from compression to tension takes place. Still, this time the non-viscous model exhibits in the case of the longitudinal RS better agreement with the measurements. A significant deviation of 100 MPa is met only at a depth between 1.5 mm and 2.5 mm. Consideration of viscosity in model *C4* this time led to no overestimation of the RS near the surface and the analysis produced quite similar results with *C3*.

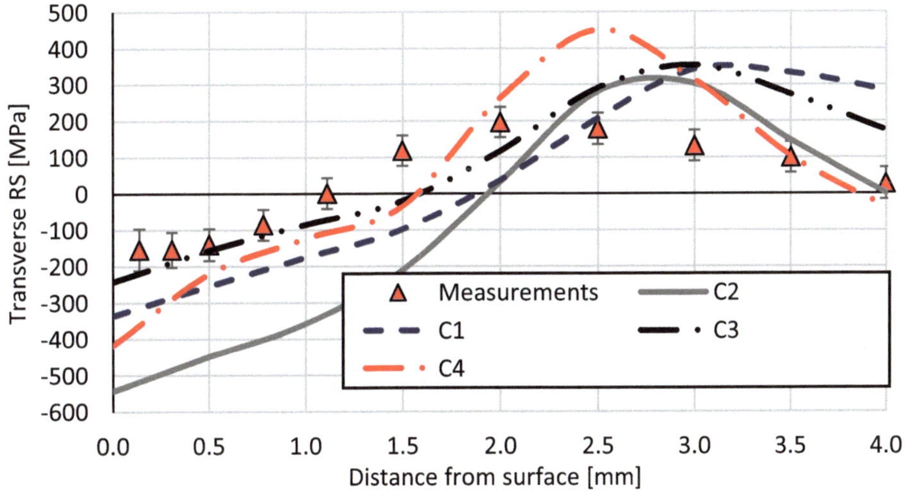

Figure 131: Simulated and measured transverse RS profiles in depth direction at the weld toe of the S960 specimen treated with a pin of 3 mm diameter, at the middle of the specimen – Measurements from [146]

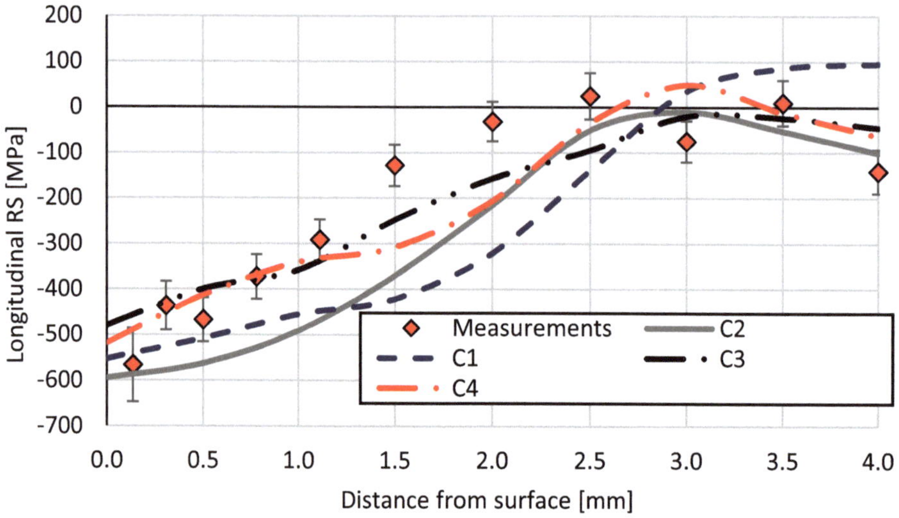

Figure 132: Simulated and measured longitudinal RS profiles in depth direction at the weld toe of the S960 specimen treated with a pin of 3 mm diameter, at the middle of the specimen – Measurements from [146]

Exemplary, the longitudinal and transverse RS contours of the model *C3* below the contact surface at a section at the middle of the investigated component and perpendicular to the introduced HFMI groove are presented in Figure 130. As in the case of the specimens of S355, the transverse RS field validates the initial expectations, with compressive stresses inside, around and underneath the HFMI groove, which are counterbalanced by tensile stresses further away from the contact surface. Moreover, the peak of RS is met as expected at a depth of around 0.8 mm underneath the contact surface. On the contrary, the peak of the longitudinal compressive RS is met on the top layers underneath the contact surface, as in the previous cases. Still, higher RS are met overall in comparison to the case of S355 due to the higher yield stress of the material. The compressive and longitudinal peak stress lie around 1000 and 1750 MPa respectively.

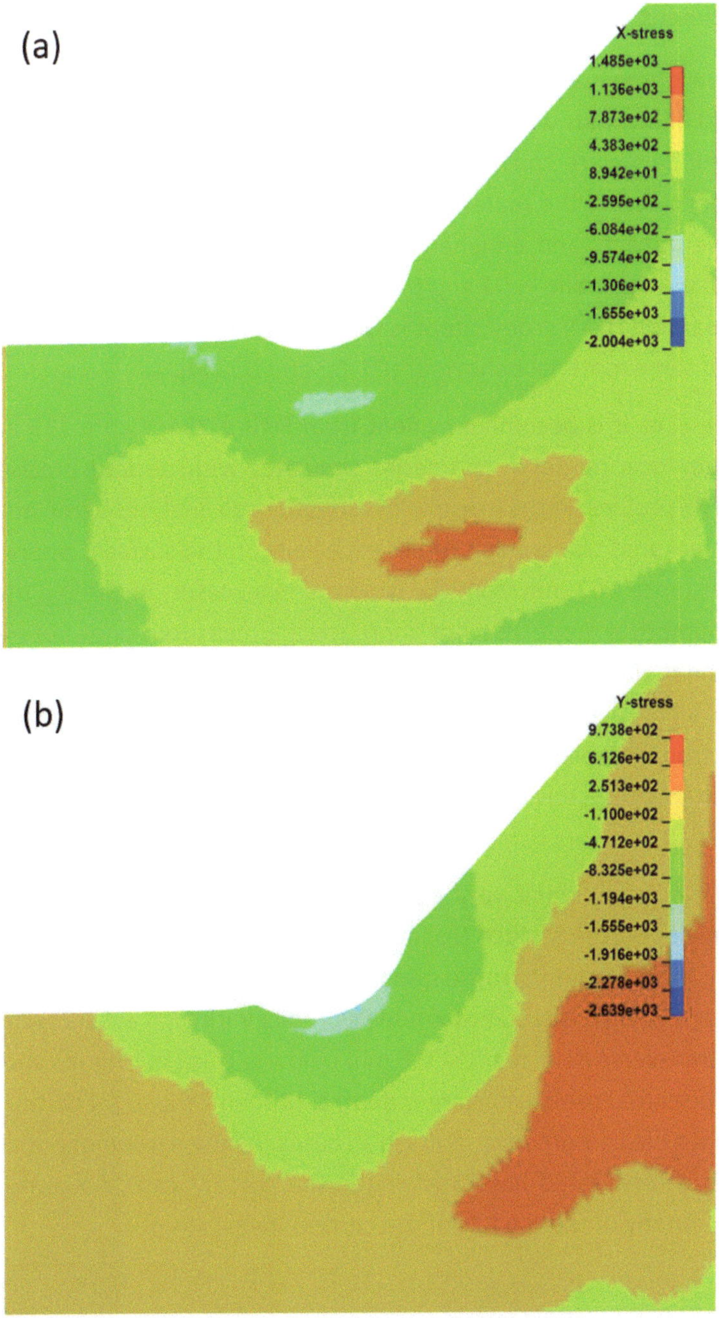

Figure 133: RS contours of the model *C2* inside and near the HFMI groove – Section at the middle of the investigated component, stresses are given in MPa: a) Transverse; b) Longitudinal

S960 Specimens Treated with a Pin of 4 mm Diameter

An overview of the numerical investigations regarding the single-pass treatments of the S960 fillet welds with a 4 mm diameter HFMI pin is given in Table 51. Four different models were considered in the present investigation, two neglecting and two considering the material's viscosity. A larger step of 0.2 mm between consecutive impacts and a higher impact velocity of 4.2 m/s were adopted in models *D1* and *D3*, while in *D2* and *D4* the respective values were 0.1 mm and 3.5 m/s. Comparable HFMI groove depths were introduced between the models that considered same material behaviour (*D1-D2* and *D3-D4*).

Nevertheless, as it is becoming evident from Figure 134, Figure 135 and Figure 136 no numerical model in this series simulates with desired preciseness the measured RS profiles. In the present investigation, all analyses agree qualitatively with the measured transverse profiles especially on the surface, where a peak is met at a distance of 3 mm from the weld toe in all cases. Regarding the longitudinal RS, numerical models do not approach the measured ones not even qualitatively. The measured profile exhibits a convex ellipsoid shape while the simulated ones a concave. Once again, numerical models overestimate the depth of transition from compression to tension both for longitudinal and transverse RS, although in this case this overestimation is more significant and almost up to 2 mm.

Models *D1* and *D3* exhibit very good quantitative agreement overall with the measurements on the surface (especially *D1*, see Figure 134) as well, excluding though the measurements at the middle of the groove, where a deviation of approximately 140 MPa is met. Both models predict with high accuracy the longitudinal and transverse RS in the first 0.5 mm below surface but significantly deviate from the measured profiles on larger depths. Models *D2 and D4* on the contrary, overestimate the compressive stress in all cases. As the agreement of all numerical models is considered non-satisfying in the present case, no contour plot is presented for any of the models of the current investigation.

Table 51: FE Analyses of the HFMI treatment on fillet welds of S960 with a pin diameter of 4 mm

Model	Material behaviour	WRS	Step [mm]	Impact velocity [m/s]	Number of passes [-]	Maximum groove depth [mm]	Vertical displacement at the weld toe [mm]
D1	non-viscous	considered	0.2	4.2	3	0.25	0.08
D2	non-viscous	considered	0.1	3.5	3	0.22	0.06
D3	viscous	considered	0.2	4.2	3	0.11	0.00
D4	viscous	considered	0.1	3.2	3	0.12	0.01

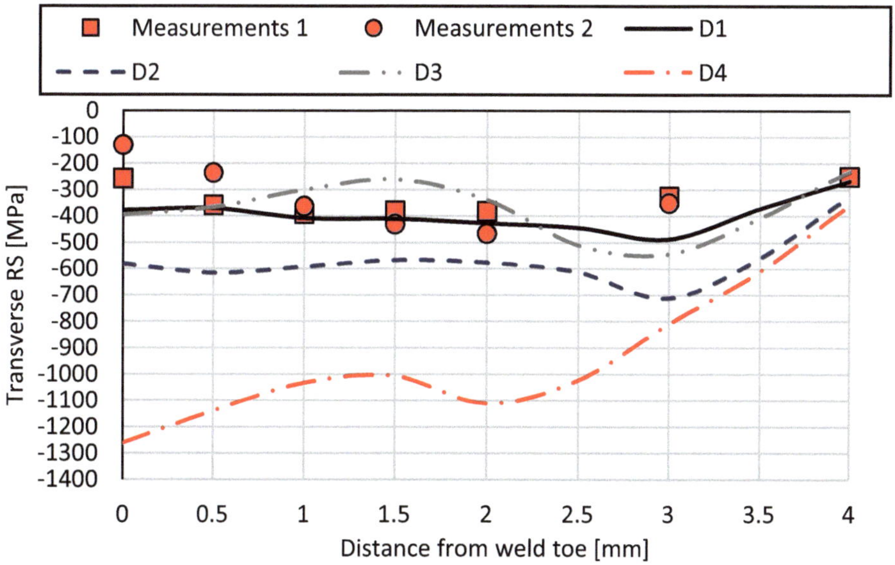

Figure 134: Simulated and measured transverse RS profiles on the surface of the S960 specimen treated with a pin of 4 mm diameter, perpendicular to the HFMI groove – Measurements from [146]

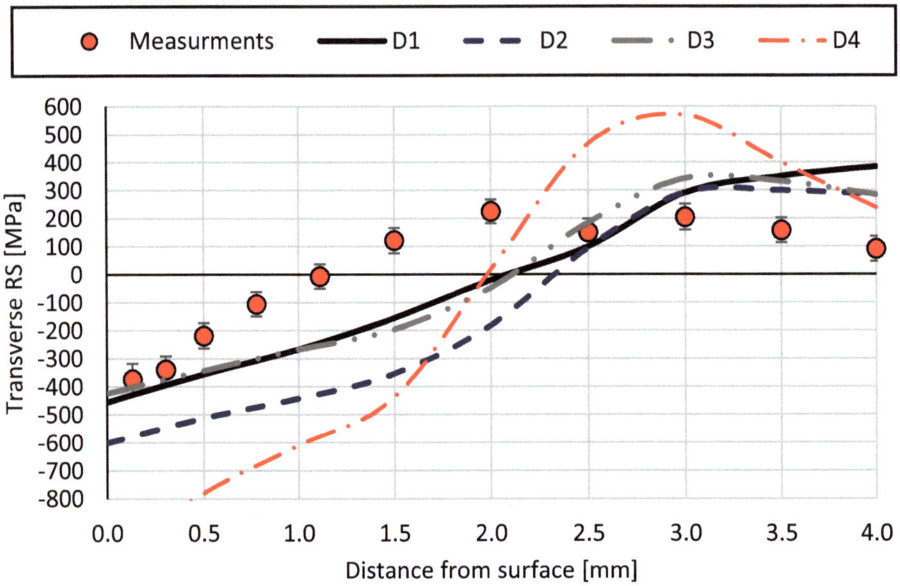

Figure 135: Simulated and measured transverse RS profiles in depth direction at the weld toe of the S960 specimen treated with a pin of 4 mm diameter, at the middle of the specimen – Measurements from [146]

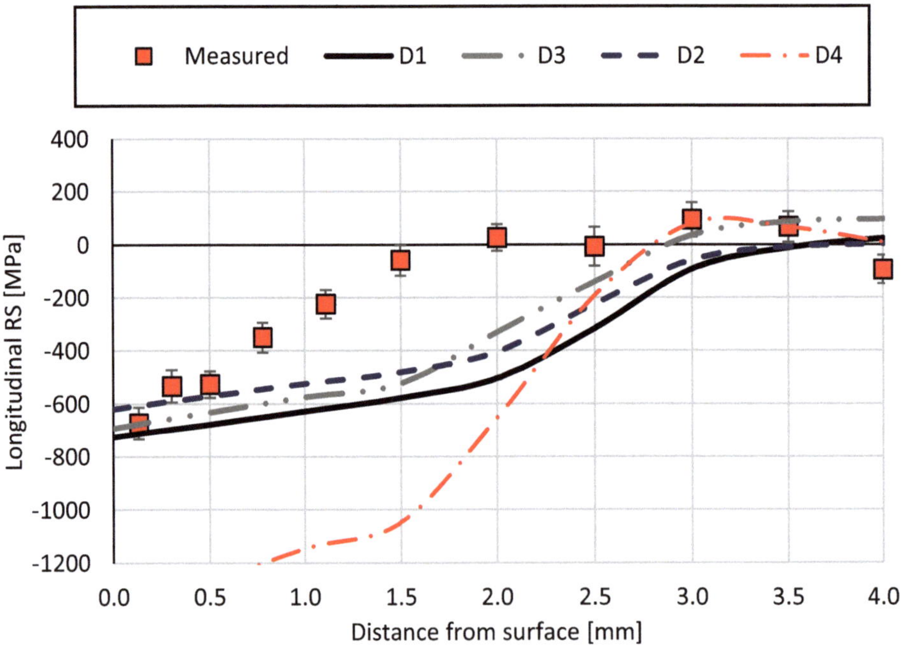

Figure 136: Simulated and measured longitudinal RS profiles in depth direction at the weld toe of the S355 specimen treated with a pin of 4 mm diameter, at the middle of the specimen – Measurements from [146]

5.4.7 Discussion and Conclusions

A series of numerical investigations regarding the HFMI treatment on the fatigue test specimens of the project *HFH-Simulation* [146] was carried out. The results of the present numerical investigations were compared with the respective measurements on the real specimens of this project. The RS were measured on the surface, perpendicular to the HFMI groove, and in depth direction, underneath the weld toe, with X-ray and neutron diffraction respectively. Details regarding the measurement can be found in [144] and [146]. Four different specimens were investigated, two of S355 and two of S960. In each case, one specimen was hammered with a pin of 3 mm and a single-pass treatment and one with a pin of 4 mm and two- or three-pass treatment, each corresponding respectively to HFMI treatments with *HiFIT* [167] and *PITec* [44]. Representative values of impact velocity taken from the measured range for each device were applied. A series of investigations was carried for each specimen in order to evaluate the significance of various aspects of the modelling approach on the preciseness of the simulated RS.

The present modelling approach predicted with sufficient accuracy in most of the above presented cases the measured RS, when WRS were considered and viscosity of the material was neglected (see models *A3, B2, C3*). Satisfying qualitative and quantitative agreement between numerically and experimentally extracted RS profiles in depth direction was met overall for the profiles apart from the case of the S960 specimen treated with a pin of 4 mm diameter. In most of the rest cases and especially for the first millimeters underneath the contact surface where fatigue cracks initiate, the deviation between measured and simulated RS was always lower than 10 % of the investigated materials yield stress, fulfilling in this case the validity criterion that was set at the beginning of the present dissertation.

Regarding the RS profiles on the surface, some cases of non-negligible discrepancies between measurements and FE analysis were found. This could be attributed to the simplified simulation of contact under the adopted Coulomb model (Eq. (28)), especially when it is considered that these discrepancies were met mostly inside the contact area. Nonetheless, this effect seems to be significantly localized on the contact surface and does not seem to influence the introduced RS field. This later statement is further supported by the above-described good agreement of simulated and measured RS profiles in depth direction, which anyway depict better the three dimensional stress state of the RS. The measured and simulated RS are averaged in this case inside the three-dimensional measured gauge volume, while for surface stresses a two-dimensional averaging area is applied.

Based on the evaluation of the above-presented investigations regarding the fillet welds the following conclusions can be drawn:

- A discretization with mesh finer than 1/40 of the pin's diameter for the present element type and for a distance of at least 1.3 x D_{pin} away from the weld toe to all directions is necessary for the precise simulation of HFMI. If contact surfaces for a transition to a volume with coarser are mesh dictated as in the present case, the outer coarse mesh should not be larger than 3.5 times the element size of the fine mesh. It has to be mentioned at this point, that small separation or penetration of the nodes on the contact surface between the volumes of fine and coarse mesh was observed in a few, individual cases without having any obvious effect on the simulated RS. No stress concentration or discontinuity was observed when the above-mentioned requirements were met. A further refinement was anyway not possible in any case due to computational restrictions.
- Modelling a travelling step of 0.2 mm between consecutive impacts, has been proven insufficient in the present case of a velocity-based simulation. Denser overlap of 0.15 mm to 0.1 mm had to be adopted depending on the materials parameters and the number of passes so that a realistic depth of the HFMI groove could be modelled under the present impact velocities. The reason for this discrepancy in comparison to the previously investigated case of parent material specimens was the local geometry of the fillet welds at the weld toe. The increased friction force in the current analyses due to the larger contact surface during the penetration of the HFMI pin to the treated material has to be accounted for the necessity of denser impacts.
- For identical material modelling different pairs of impact velocity and travelling step may have introduced grooves of approximately similar depths but the simulated transverse RS profiles deviate with each other significantly (see for example models *D1* and *D2*). Therefore, the density of the impacts exhibits a significant influence on the modelled transverse RS as it was proposed earlier in [145]. Hence, a limitation to the assumption of a

step larger than the real for the reduction of the computational time, a simplification that was initially based on the experimental results from [44], is met as expected.

- Comparing analyses, which assumed identical non-viscous material properties and overlap of impacts (*B1* with *B2 or B3* with *B4 etc.*), the significance of modelling with preciseness the depth and in extension the geometry of the HFMI groove, is becoming evident. When no further information is available, modelling a HFMI groove of 0.15 mm for S355, i.e. the mean value of the range (0.10 mm - 0.20 mm) proposed by IIW Recommendations for the quality assurance of the HFMI treatment [118], and a groove of 0.125 mm for S960 in order to compensate for the harder material is recommended for practical applications.
- The significant deviation of the simulated RS in cases between pairs of identical models, which either neglected or considered the WRS (*A1* and *A3* etc.) and the significantly better results of the latter exhibit the significance of considering WRS during a HFMI simulation. Ergo, any assumption that WRS and the respective plastic strains are negligible in comparison to those introduced by HFMI and they should not be considered during its modelling is not valid.
- Similarly to the case of modelling the HFMI treatment of parent material specimens of S690 and S960, the consideration of the material's viscosity in the present investigations led to overestimation of the compressive RS and a significant deviation between modelled and measured RS profiles. The previously stated assumption that for the high strength materials the strain rate dependency can be lower under the present deformation mode than that extracted by tensile tests at high strain rates is emphasized again at this point. It has to be underlined that even for the present S355 specimens, the higher yield stress of the HAZ was considered which is at the same order of magnitude with that of the parent material S690 (713 MPa and 734 MPa). Nonetheless, this incompatibility could be attributed to the simplicity of the applied Cowper-Symonds model (Eq. (12)), which only scales the yield stress based on the plastic strain rate and not the hardening behaviour as well. It is possible, that with a more elaborate material model (Chaboche model [22], see [143] for instance) an accurate simulation of RS with consideration of the materials viscocity could be possible. Still, the good agreement of the

current analyses considering a non-viscous model, when the impact density and velocity and the WRS are modelled accurately, constitutes the respective models valid and allows for the neglection of strain rate dependency, at least for the currently investigated materials and especially for practical applications.

Based on the evaluation of the above-presented investigations, which satisfied the accuracy criteria of the present study and are considered valid, i.e. the non-viscous material models with consideration of WRS and accurate simulation of the HFMI groove the following conclusions were drawn:

- The RS contours bellow the HFMI groove validate the theoretical distributions. The stresses around and near the HFMI trace are compressive and they are counterbalanced by outer tensile stresses. The peak compressive transverse stresses are met in an elliptical area less than a millimetre deeper underneath the treatment surface and they are approximately equal to the yield stress of the material under treatment, i.e. the yield stress of the HAZ. On the contrary, peak longitudinal RS are met on the top layer elements underneath the treatment surface and they are significantly higher than the HAZ yield stress (80 % to 100 % higher). These peaks could not be depicted by the measurements in the depth direction due to the large gauge volumes, which were applied during measurements (2 mm x 2 mm x 2 mm and 5 mm x 2 mm x 2 mm for longitudinal and transverse direction respectively). As the average of the RS inside these volumes is extracted at each measurement point, the finer fluctuations of the RS profiles that can be seen on the RS contour plots could not be measured.
- The above-stated assumption regarding the scaling-down of the component during FE simulation of the HFMI treatment and the proposed requirements regarding the modelled HFMI treatment length and the distance from the edges of the component are validated. The initial assumption that the local nature of the HFMI treatment allows for the precise simulation of the RS field of a large HFMI-treated component on a small numerical model stands and is proposed for practical applications.
- The current valid models show better agreement in the case of a single-pass treatment than in the case of multiple treatment passes (*A3, C3* vs *B2*), although sufficient accuracy was met in all cases for the purpose of the present

study. The influence of cyclic plasticity is accounted for the reduced preciseness in the latter case. As it was proposed above for the consideration of viscosity, further improvement could be achieved by adoption of more elaborate material modelling such as multilinear σ-ε constitutive relation and mixed hardening behaviour. The last one has already been proven to provide more accurate results than kinematic hardening in previous studies (see [45], [127]).

Finally, the deviation in some cases between the two series of measurements, which were carried out on the surface of the specimens, reveals the complexity of the subject of RS. These measurements, which were carried out parallel to each other at the centre of the specimens and only some mm away from each other on macroscopically identical locations, should be theoretically under the same stress state. Still, they exhibit deviations of up to 10 % of the materials yield limit. This observation on the one side validates the suitability of the thumb rule of 10 % agreement between measurements and FE analyses of RS as a criterion for the validity of the latter one. On the other side, it is becoming evident that local effects which cannot be easily included in a FE analysis, such as stress concentration due to local geometrical anomalies or localized changes of hardness, can have a small effect of this order of magnitude on the RS state.

5.5 Summary and Open Questions Regarding HFMI Simulation

A series of FE analyses regarding the HFMI treatment of parent material and welded specimens were carried out in the framework of the present dissertation's last chapter. In the case of parent material specimens, steels S355, S690 and S960 were investigated. In the case of the fillet welds, only the first and the last ones were investigated. In both cases, measurements of RS were applied for the validation of the results based on the thumb rule of 10 %, as it was set at the beginning of the current study. Two different approaches for modelling the boundary conditions regarding the movement of the HFMI pin were investigated, a displacement- and a velocity-based. A new, more precise evaluation of a previous displacement-based study of the author have shown that this method as well can provide sufficient results. Nonetheless, the velocity-based approach was preferred in most cases, as it lies closer to physical reality.

During the validation of the present modelling approach, significant conclusions were drawn regarding the predominant aspects of the HFMI modelling. In most cases, neglecting the strain-rate dependency of the material in the FE analyses led to a very good agreement between modelled and measured RS profiles. Introducing viscosity led to increased accuracy only in the simulation of the parent material S355, the only case where the treated material would not be classified as high strength steel. On the contrary, it has been shown that during modelling of HFMI treatment on welded components, WRS should be considered as input parameter. Otherwise, significant overestimation of the compressive RS field after completion of the treatment can take place and therefore, contribute to a non-conservative estimation of the investigated components fatigue life.

Further investigations on an even more elaborate material modelling, which could provide an even better agreement between measured and simulated RS, is strongly recommended in future work [143]. Finally, the present validated modelling approach should be applied as a base for a sensitivity analysis regarding the setup parameters of the HFMI treatment such as HFMI groove depth, impact velocity, travelling velocity, pin diameter, angle of application etc. in an effort to maximize the magnitude of the introduced compressive RS and in extension the positive effect of the HFMI treatment on the fatigue life of the welded components.

6 Overall Discussion

Goal of the present doctoral dissertation was the investigation of the possibilities regarding the estimation of RS introduced by welding and the HFMI treatment by means of FE analysis. Initial intention was the establishment of a validated approach, which would estimate with sufficient preciseness the RS field, taking into consideration all the predominant aspects that would have a significant influence. Nevertheless, it was considered fundamental that the established approach could be reproduced in practice by a structural engineer with only basic knowledge of material science if possible. In order to fulfil this latter condition, it was required by the author to investigate in depth not only practical but special aspects of the simulation, such as elaborate material modelling or boundary conditions modelling influence of numerical effects, behaviour of various materials etc. Moreover, during the realization of the present dissertation, various relevant aspects regarding the investigated subject came up, which were not included in the initial working plan and were necessary for the fulfilment of the research goal. It was believed though that their investigation would widen the current knowledge on the field of RS simulation.

A straightforward engineering modelling approach was developed and proposed for the consideration of microstructural transformations during welding. Sophisticated aspects of material science like phase transformations or TRIP were taken into consideration, based on new or existing practical solutions, without changing the FE formulations. The level of the method's preciseness can be adjusted depending on the accuracy desired by the modeller in each investigated case. Under its current form the approach provided significantly better agreement with measured RS profiles than previous models, fulfilling the set goal of the present dissertation. The established method can act either as a validation base for existing measured welds or predictively for new ones and is applicable for other materials as well.

Similarly and based on the same principles regarding the applied methodology of the complex investigated subjects and the straightforwardness of the final established model, an approach for the simulation of HFMI was developed as well. A good agreement between measured and simulated RS profiles with the present approach was met once again. The influence of several simulation's aspects on the modelled RS profiles was investigated and conclusions, which can act as a guide for practical approach, were drawn.

For the calibration of the material models, which were applied in the numerical investigations, a series of drop tests was conceived and implemented for the estimation of the material's behaviour under the deformation mode introduced by HFMI. Useful conclusions were drawn and the method seems promising, although difficulties regarding the evaluation of the results were encountered.

Future work on the individual parts of the present study has been proposed in the respective chapters. In the next, final chapter of the current manuscript, possibilities regarding the future application of the present approach in the framework of a holistic estimation of a HFMI-treated weldment's fatigue life are discussed.

7 Future Work on Numerical Investigations and Fatigue

If fatigue of steel and metallic components in general could be investigated in a holistic approach it would constitute a multidisciplinary subject. Of course, fatigue tests accommodated by appropriate, modern measurements and monitoring enable not only a clear overview of the total fatigue life but precise information about the regimes of crack initiation and growth. Other than in material science, in civil engineering the describing models remain in most cases phenomenological and are calibrated based on that data. On the contrary, validated numerical analyses can enable a detailed description of the physical phenomena themselves. Numerical investigations regarding the prediction of fatigue life can be tracked in studies from different subjects of modern science though. Suitable FE analyses can predict the crack initiation at metallic crystalline grain level and they are already applied in the field of material science (see [14] for instance). In the state-of-the-art investigations of mechanical components, a damage tolerant approach is adopted and an estimation of the propagation of existing fatigue cracks is carried out based on fracture mechanics (see for instance [123] or [129]). Such investigations are common in modern mechanical engineering. Auxiliary to this fracture mechanics models, the evolution of the WRS simulation methods in the last years allows for consideration of the stress fields on the investigated crack's tip, which can increase significantly their preciseness. Similarly, the present study shows that a simulation of the RS field introduced by HFMI is possible. Until today though, the restricted computational capacity has been the main reason for breaking down the phenomena of fatigue into different regimes. Still, the increase in computational power will possibly allow in the upcoming years the synergy of the above-mentioned numerical investigations, up to the composition of a holistic simulation approach. Such an approach could predict the full fatigue life of a real scale component from the crack initiation at the grain level up to fracture. The use of sub-modelling techniques could further accelerate this composition. Therewith, an optimal design for each unique mechanical component against fatigue would be enabled leading to a significant upgrade of engineering efficiency.

References

[1] Acevedo C.; *Influence of Residual Stresses on Fatigue Response of Welded Tubular K-Joints*, Doctoral Thesis, École Polytechnique Fédérale de Lausanne 2011.

[2] Acevedo C., Drezet J. M., Nussbaumer A.; Numerical Modelling and Experimental Investigation on Welding Residual Stresses in Large-Scale Tubular K-Joints, *Fatigue & Fracture of Engineering Materials & Structures* **36**, pp. 177–185, 2012.

[3] Andersson B.; Thermal Stresses in a Submerged-Arc Welded Joint Considering Phase Transformations, *Transactions of the ASME* **100**, pp. 356–362, 1978.

[4] ANSYS® Academic Research, Release 19.2, Help System, ANSYS, Inc., 2018.

[5] Argyris J. H., Sjimmat J., Willam K. J.; Computational Aspects of Welding Stress Analysis, *Computer Methods in Applied Mechanics and Engineering* **33** (1–3), pp. 635-665, 1982.

[6] Avallone E. A., Baumeister T.; *Mark's Handbookf for Mechanical Engineers*, 10th Edition, McGraw-Hill, New York, 1996.

[7] Avrami M.; Granulation, Phase Change and Microstructures - Kinetics of Phase Change. III, *Journal of Chemical Physics* **9**, pp. 177-184, 1941.

[8] Avramov I., Sestak J.; Generalized Kinetics of Overall Phase Transition Explicit to Crystallization, *Journal of Thermal Analysis and Calorimetry* **118**, pp. 1715-1720, 2014.

[9] Bai W., Ronglei S., Leopold J.; Numerical Modelling of Microstructure Evolution in Ti6Al4V Alloy by Ultrasonic Assisted Cutting, *7th HPC 2016 – CIRP Conference on High Performance Cutting, Procedia CIRP* **46**, pp. 428 – 431, 2016.

[10] Baptista R., Infante V., Branco C.; Fully Dynamic Numerical Simulation of the Hammer Peening Fatigue Life Improvement Technique, *Procedia Engineering* **10**, pp. 1943-1948, 2011.

[11] Bhatti A. A., Barsoum Z., Murakawa H., Barsoum I.; Influence of Thermo-Mechanical Material Properties of Different Steel Grades on Welding Residual Stresses and Angular Distortion, *Materials and Design* **65,** pp. 878–889, 2015.

[12] Blacha S., Weglowski M. S., Dymek S., Kopuscianski, M.; Microstructural Characterization and Mechanical Properties of Electron Beam Welded Joint of High Strength Steel Grade S690QL, *Archives Of Metallurgy And Materials* **61** (2B), pp. 1193–1200, 2016.

[13] Bodner S. R., Symonds P. S.; Experimental and Theoretical Investigation of the Plastic Deformation of Cantilever Beams Subjected to Impulsive Loading, *Journal of Applied Mechanics* **29**, pp. 719–728, 1962.

[14] Boeff M., Hassan H. U., Hartmaier A.; Micromechanical Modelling of Fatigue Crack Initiation in Polycrystals, *Journal of Material Research* **32** (23), pp. 4375-4386, 2017.

[15] Börjesson L., Lindgren L.-E.; Simulation of Multipass Welding With Simultaneous Computation of Material Properties, *Journal of Engineering Materials and Technology* **123**, pp. 106-111, 2001.

[16] Boyce B. L., Chen X., Hutchinson J. W., Ritchie R. O.; The Residual Stress State due to a Spherical Hard-Body Impact, *Mechanics of Materials* **33**, pp. 441-454, 2001.

[17] Brickstad B., Josefson B. L.; A Parametric Study of Residual Stresses in Multi-Pass Butt-Welded Stainless Steel Pipes, *International Journal of Pressure Vessels and Piping* **75**, pp. 11-25, 1998.

[18] Brown S., Song H.; Finite Element Simulation of Welding of Large Structures, *Journal of Engineering for Industry* **114**, pp. 440-451, 1992.

[19] Byfield M. P., Davies J. M., Dhanalakshmi M.; Calculation of the Strain Hardening Behaviour of Steel Structures Based on Mill Tests, *Journal of Constructional Steel Research* **61**, pp. 133-150, 2005.

[20] Cadoni E., Forni D., Gieleta R., Kruszka L.; Tensile and Compressive Behaviour of S355 Mild Steel in a Wide Range of Strain Rates; *The European Physical Journal Special Topics* **227**, pp. 29-43, 2018.

[21] Caron J., Heinze C., Schwenk C., Rethmeier M., Babu S. S., Lippold J.; Effect of Continuous Cooling Transformation Variations on Numerical Calculation of Welding-Induced Residual Stresses, *Welding Journal* **89**, pp. 151-160, 2010.

[22] Chaboche J. L.; Constitutive Equations for Cycling Plasticity and Cyclic Viscoplasticity, *International Journal of Plasticity* **5**, pp. 247-302, 1989.

[23] Choobi M. S.; Investigating the Effect of Geometrical Parameters on Distortions in Butt-Welded Plates, *The Journal of Strain Analysis for Engineering Design* **48** (4), pp. 258–268, 2013.

[24] Chopra A. K.; *Dynamics of Structures – Theory and Application to Earthquake Engineering*, 4th Edition, Pearson Prentice Hall, New Jersey 2012.

[25] Cowper G. R., Symonds P. S.; *Strain Hardening and Strain rate Effects in the Impact Loading of Cantilever Beams*, Report No. 28, Brown University Division of Applied Mathematics, September 1957.

[26] Dattoma V., De Giorgi M., Nobile R.; On the Evolution of Welding Residual Stress after Milling and Cutting Machining, *Computers and Structures* **84**, pp. 1965–1976, 2006.

[27] De Benardos N., Olszewski S; *Process of AMD Apparatus for Working Metals by the Direct Application of the Electric Current*, No. 363, 320, 1887.

[28] Deng D., Murakawa H.; Numerical Simulation of Temperature Field and Residual Stress in Multi-Pass Welds in Stainless Steel Pipe and Comparison with Experimental Measurements, *Computational Materials Science* **37,** pp. 269–277, 2006.

[29] DIN EN 10025-2:2005-04; *Hot Rolled Products of Structural Steels - Part 2: Technical Delivery Conditions for Non-Alloy Structural Steels*, 2004.

[30] DIN EN 10025-6:2018-07; *Hot Rolled Products of Structural Steels - Part 6: Technical Delivery Conditions for Flat Products of High Yield Strength Structural Steels in the Quenched and Tempered Conditions*, 2018.

[31] Dong P.; Residual Stress Analyses of a Multi-Pass Girth Weld: 3-D Special Shell versus Axisymmetric Models, *Journal of Pressure Vessel Technology* **123**, pp. 207-213, 2001.

[32] Dupont J. N., Marder A. R.; Thermal Efficiency of Arc Welding Processes, *Welding Journal* **74,** pp. 406–416, 1995.

[33] Dürr A.; *Zur Ermüdungsfestigkeit von Schweißkonstruktionen aus höherfesten Baustählen bei Anwendung von UIT-Nachbehandlung,* Dissertation, Fakultät Bau- und Umweltingenieurwissenschaften, Universität Stuttgart 2006.

[34] EN 1993-1-9; *Eurocode 3: Design of Steel Structures - Part 1-2: General Rules - Structural Fire Design*, 2009.

[35] EN 1993-1-9; *Eurocode 3 - Design of Steel Structures - Part 1-9: Fatigue*, 2009.

[36] Ernould C., Schubnell J., Farajian M., Maciolek A., Simunek D., Leitner M., Stotschka M.; Applications of Different Simulation Approaches to

Numerically Optimize High-Frequency Mechanical Impact (HFMI) Post-Treatment Process, *Welding in the World* **63** (3), pp. 725–738, 2019.

[37] DIN EN 573-3:2018-12; *Aluminium and Aluminium Aalloys - Chemical Composition and Form of Wrought Products - Part 3: Chemical Composition and Form of Products*, 2018

[38] Feldmann M., Bartsch H., Ummenhofer T., Seyfried B., Kuhlmann U., Drebenstedt K.; *Neubewertung und Erweiterung des Kerbfallkataloges nach Eurocode 3 für eine zukunftsfähige Auslegung hochbeanspruchter Stahlkonstruktionen*, Schlussbericht, im Auftrag von Deutscher Ausschuss für Stahlbau (DASt) in Kooperation mit Forschungsvereinigung Stahlanwendung e.V. FOSTA, IGF 19178 N, 2019.

[39] Ferro P., Bonollo F.; Modelling of the Carburizing and Quenching Process Applied to Caterpillar Track Bushings, *Modelling and Simulation in Materials Science and Engineering* **22** (2), 2014.

[40] Ferro P., Porzner H., Tiziani A., Bonollo F.; The Influence of Phase Transformations on Residual Stresses Induced by the Welding Process—3D and 2D Numerical Models, *Modelling and Simulation in Materials Science and Engineering* **14**, pp. 117–136, 2006.

[41] Fischer-Cripps A. C.; *Introduction to Contact Mechanics*, 2nd Edition, Springer, Boston, 2007.

[42] Fischer F. D., Sun Q.-P., Tanaka K.; Transformation-Induced Plasticity (TRIP), *ASME Appied Mechanics Reviews* **49** (6), pp. 317-364, 1996.

[43] Flanagan D. P., Belytcshko T.; A Uniform Strain Hexaedron and Quadrilateral with Orthogonal Hourglass Control, *International Journal for Numerical Methods in Engineering* **17**, pp. 679-706, 1981.

[44] Foehrenbach J., Hardenacke V., Farajian M.; High Frequency Mechanical Impact Treatment (HFMI) for the Fatigue Improvement: Numerical and Experimental Investigations to describe the Condition in the Surface Layer, *Welding in the World* **60** (4), pp. 749–755, 2016.

[45] Foehrenbach J., Hardenacke V., Farajian M., Luke M.; *Fatigue Life Calculation of High Frequency Mechanical Impact (HFMI) Treated Welded Joints by means of Numerical Process Simulation and Critical Plane Approaches*, IIW Commission XIII, Doc. XIII-2637-16, 2016.

[46] Forni D., Chiaia B., Cadoni E.; Blast Effects on Steel Columns under Fire Conditions, *Journal of Constructional Steel Research* **136**, pp. 1-10, 2017.

[47] Forni D., Chiaia B., Cadoni E.; High Strain Rate Response of S355 at High Temperatures, *Materials and Design* **94**, pp. 467-478, 2016.

[48] Forni D., Chiaia B., Cadoni E.; Mechanical Properties of S355 under Extreme Coupled Effect of High Temperatures and High Strain Rates, *ECCOMAS Congress 2016 VII European Congress on Computational Methods in Applied Sciences and Engineering M. Papadrakakis, V. Papadopoulos, G. Stefanou, V. Plevris (eds.),* Crete Island 5-10 June 2016.

[49] Forni D., Chiaia B., Cadoni E.; Strain Rate Behaviour in Tension of S355 Steel: Base for Progressive Collapse Analysis, *Engineering Structures* **119**, pp. 167-173, 2016.

[50] Forrestal M. J., Sagartz M. J.; Elastic–Plastic Response of 304 Stainless Steel Beams to Impulse Loads, *Journal of Applied Mechanics* **45**, pp. 685–687, 1978.

[51] Francis J. A., Bhadeshia H. K. D. H., Withers P. J.; Welding Residual Stresses in Ferritic Power Plant Steels, *Material Science and Technology* **23**, pp. 1009-1020, 2007.

[52] Fujita Y., Takesshi Y., Kitamura M., Nomoto T.; *Welding Stresses with Special Reference to Cracking,* IIW Doc X-655-72, 1972.

[53] Gkatzogiannis S., Knoedel P., Ummenhofer T.; A Pseudothermal Approach for Simulating the Residual Stress Field Caused by Shot Blasting, *Proceedings of the VIII International Conference on Computational Methods for Coupled Problems in Science and Engineering, Sitges 3-6 June 2019,* pp. 777-788, 2019.

[54] Gkatzogiannis S., Knoedel P., Ummenhofer T.; FE Simulation of High Frequency Mechanical Impact (HFMI) Treatment – First Results, *Proceeding of NORDIC STEEL 2019, The 14th Nordic Steel Construction Conference, Copenhagen 18–20 September 2019, ce/papers* **3** (3-4), pp. 797-802, Ernst & Sohn, 2019.

[55] Gkatzogiannis S., Knoedel P., Ummenhofer T.; FE Welding Residual Stress Simulation – Influence of Boundary Conditions and Material Models, *Proceedings of EUROSTEEL 2017, Copenhagen 13–15 September 2017, ce/papers* **1**, Ernst & Sohn, 2017.

[56] Gkatzogiannis S., Knoedel P., Ummenhofer T.; Influence of Welding Parameters on the Welding Residual Stresses, *Proceedings of the VII International Conference on Coupled Problems in Science and Engineering, Rhodes 12–14 June 2017*, pp. 767–778, 2017.

[57] Gkatzogiannis S., Knoedel P., Ummenhofer T.; Reviewing the Influence of Welding Setup on FE-simulated Welding Residual Stresses, *Proceedings of the 10th European Conference on Residual Stresses - ECRS10, Leuven 11-14 September 2018,* published in *Materials Research Proceedings* **6**, pp. 197-202, 2018.

[58] Gkatzogiannis S., Knoedel P., Ummenhofer T.; Simulation of Welding Residual Stresses – From Theory to Practice, Selected Peer Reviewed Papers from the *12th International Seminar Numerical Analysis of Weldability, Graz – Schloss Seggau 23-26 September,* published in *Sommitsch C., Enzinger N., Mayr P.; Mathematical Modelling of Weld Phenomena* **12**, pp. 383-400, 2019.

[59] Gkatzogiannis S., Knoedel P., Ummenhofer T.; Strain Rate Dependency of Simulated Welding Residual Stresses, *Journal of Material Engineering and Performance* **27** (10), pp. 5079-5085, 2017.

[60] Goldak J. A., Akhlaghi M.; *Computational Welding Mechanics*, 1st Edition, Springer US, New York, 2005.

[61] Goldak J. A., Chakravarti A., Bibby M.; A New Finite Element Model for Welding Heat Sources, *Metallurgical Transactions B* **15,** pp. 299–305, 1984.

[62] Greenwood G. W., Johnson R. H.; The Deformation of Metals Under Small Stresses During Phase Transformation, *Proceedings of the Royal Society A* **283,** pp. 403-422, 1965.

[63] Guagliano M.; Relating Almen Intensity to Residual Stresses Induced by Shot Peening: a Numerical Approach, *Journal of Materials Processing Technology* **110**, pp. 277-286, 2001.

[64] Guo C., Zhijiang W., Dongpo W., Shengsun H.; Numerical Analysis of the Residual Stress in Ultrasonic Impact Treatment Process with Single-Impact and Two-Impact Mode, *Applied Surface Science* **347**, pp. 596-601, 2015.

[65] Gupta C., Chakravartty J. K., Wadekar S. L., Dubey J. S.; Effect of Serrated Flow on Deformation Behaviour of AISI 403 Stainless Steel, *Materials Science and Engineering A* **292**, pp. 49-55, 2000.

[66] Haidemenopoulos G.; *Εισαγωγή στις Συγκολλήσεις* [Introduction to Welding], 1st Edition, Tziolas Publications, Thessaloniki 2010.

[67] Haidemenopoulos G.; *Φυσική Μετταλουργία* [Physical Metallurgy], 1st Edition, Tziolas Publications, Thessaloniki 2007.

[68] Hardenacke V., Farajian M., Siegele D.; *Modelling and Simulation of the High Frequency Mechanical Impact (HFMI) Treatment of Welded Joints*, IIW Commission XIII, *Doc. XIII-2533-14*, 2014.

[69] Heinze C., Schwenk C., Rethmeier M., Caron J.; Numerical Sensitivity Analysis of Welding-Induced Residual Stress Depending on Variations in Continuous Cooling Transformation Behavior, *Frontiers of Material Science* **5** (2), pp. 168–178, 2011.

[70] Heinze C., Schwenk C., Rethmeier M.; Numerical Calculation of Residual Stress Development of Multi-Pass Gas Metal Arc Welding, *Journal of Constructional Steel Research* **72**, pp. 12-19, 2012.

[71] Hensel J., Nitschke-Pagel T., Dilger K.; On The Effects of Austenite Phase Transformation on Welding Residual Stresses in Non-Load Carrying Longitudinal Welds, *Welding in the World* **59**, pp. 179–190, 2015.

[72] Hensel J., Nitschke-Pagel T., Eslami H., Dilger K.; Fatigue Strength Enhancement of Butt Welds by Means of Shot Peening and Clean Blasting, *IIW Document XIII-2733-18*, 2018.

[73] Hertz H.; Über die Berührung fester elastischer Körper, J. *Reine und Angewandte Mathematik* **92**, pp. 156-171, 1881.

[74] Hildebrand J.; *Numerische Schweißsimulation: Bestimmung von Temperatur, Gefüge und Eigenspannung an Schweißverbindungen aus Stahl- und Glaswerkstoffen*, Dissertation, Fakultät Bauingenieurwesen, Bauhaus-Universität Weimar 2008.

[75] Hill R.; *The Mathematical Theory of Plasticity*, 1st Edition, Oxford University Press, New York, 1950.

[76] Hobbacher A.; *Recommendations for Fatigue Design of Welded Joints and Components*, IIW Commission XIII, Doc. XIII-2151r4-07/XV-1254r4-07, 2008.

[77] Hong T., Ooi J. Y., Shaw B.; A Numerical Simulation to Relate the Shot peening parameters to the Induced Residual Stresses, *Engineering Failure Analysis* **15**, pp. 1097–1110, 2008.

[78] Huh H., Lim J. H., Song J. H., Lee K.-S., Lee Y.-W., Han S. S.; Crashworthiness Assessment of Side Impact of an Auto-Body with 60trip Steel for Side Members, *International Journal of Automotive Technology* **4** (3), pp. 146-156, 2003.

[79] ISO 5817:2014; *Welding - Fusion-Welded Joints in Steel, Nickel, Titanium and Their Alloys (Beam Welding Excluded) - Quality Levels for Imperfections*, 3rd Edition.

[80] ISO 19675:2017; *Non-destructive testing — Ultrasonic testing — Specification for a calibration block for phased array testing (PAUT)*, 2017.

[81] Jackson R, Green I.; A Finite Element Study of Elasto-plastic Hemispherical Contact, *Proceedings of 2003 STLE/ASME Joint International Tribology Conference,* Ponte Vedra Beach 26-29 October 2003, .

[82] Johnson K. L.; *Contact Mechanics,* 1st Edition, Cambridge University Press, Cambridge (UK), 1985.

[83] Johnson K. L.; The Correlation of Indentation Experiments, *Journal of the Mechanics and Physics of Solids* **18**, pp. 115-126, 1970.

[84] Johnson G. R., Cook W. H.; A Constitutive Model and Data for Metals subjected to Larger Strains, High Strain Rates and High Temperatures, *Proceedings of the 7th International Symposium on Ballistics, the Hague 19-21 April 1983*, pp. 541–547, 1983.

[85] Johnson W., Mehl R.; Reaction Kinetics in Processes of Nucleation and Growth, *Transactions of American Institute of Mining and Metallurgical Engineers* **135**, pp. 416-458, 1939.

[86] Jones, N.; *Structural Impact,* 2nd Edition, Cambridge University Press, New York, 2012.

[87] Karlsson C. T.; Finite Element Analysis of Temperatures and Stresses in a Single-Pass Butt-Welded Pipe — Influence Of Mesh Density And Material Modelling, *Engineering with Computers* **6**, pp. 133–141, 1989.

[88] Katsareas D. E., Ohms C., Youtsos A. G.; Finite Element Simulation of Welding in Pipes: A Sensitivity Analysis, *Proceedings of a Special Symposium held within the 16th European Conference of Fracture – ECF16, Alexandroupolis, 3-7 July 2006*, pp. 15-26, 2006.

[89] Kim T., Lee H., Hyun H. C., Jung S.; A Simple but Effective FE Model with Plastic Shot for Evaluation of Peening Residual Stress and its Experimental Validation, *Science and Engineering A* **528**, pp. 5945–5954, 2011.

[90] Kim T., Lee H., Jung S., Lee J. H.; A 3D FE Model with Plastic Shot for Evaluation of Equi-Biaxial Peening Residual Stress due to Multi-Impacts, *Surface & Coatings Technology* **206**, pp. 3125–3136, 2012.

[91] Knobloch M., Pauli J., Fontana M.; Influence of the Strain Rate on the Mechanical Properties of Mild Carbon Steel at Elevated Temperatures, *Materials & Design* **49**, pp. 553-565, 2013.

[92] Knoedel P., Gkatzogiannis, S., Ummenhofer T.; Practical Aspects of Welding Residual Stress Simulation, Journal of Constructional Steel Research **132**, pp. 83-96, 2017.

[93] Knoedel P., Gkatzogiannis S., Ummenhofer T.; FE Simulation of Residual Welding Stresses: Aluminum and Steel Structural Components, *Key Engineering Materials* **710**, pp. 268-274, 2016.

[94] Koistinen D. P., Marburger R. E.; A General Equation Prescribing the Extent of the Austenite-Martensite Transformation in Pure Iron-Carbon Alloys and Plain Carbon Steels, *Acta Metall.* **7,** pp. 59–61, 1959.

[95] Kolmogorov A.; Statistical Theory of Crystallization of Metals (in Russian), *Izvestia Akameia Nauk SSSR Ser. Mathematica* **1**(1), pp. 355-359, 1937 (cited after [111]).

[96] Kou S.; *Welding Metallurgy*, 2nd Edition, John Wiley & Sons, Hoboken New Jersey, 2003.

[97] Kraus H. G.; Experimental Measurement of Thin Plate 304 Stainless Steel GTA Weld Pool Surface Temperatures, *Welding Research Supplement*, pp. 353-359, 1987.

[98] Kuhlmann U.; *Experimentelle Untersuchungen zur ermüdungssteigernden Wirkung des PIT-Verfahrens*, Versuchsbericht, Fakultät Bau- und Umweltingenieurwissenschaften, Universität Stuttgart 2009.

[99] Kuilin Y., Sumi Y.; Simulation of Residual Stress and Fatigue Strength of Welded Joints Under the Effects of Ultrasonic Impact Treatment (UIT), *International Journal of Fatigue* **92,** pp. 321-332, 2016.

[100] Kuilin Y., Sumi Y.; Modelling of Ultrasonic Impact Treatment (UIT) of Welded Joints and its Effect on Fatigue Strength, *Frattura ed Integrità Strutturale* **34,** pp. 476-486, 2015.

[101] Kurtara H., Buyuk M., Eskandarian A.; Ballistic Impact Simulation of GT Model Vehicle Door Using Finite Element Method, *Theoretical and Applied Fracture Mechanics* **40**, pp. 113-121, 2003.

[102] Ladendorf P., Schubnell J., Knoedel P., Farajian M., Ummenhofer T., Luke T.; *Entwicklung einer verifizierten Prozedur für die zuverlässige schweißtechnische Instandsetzung von Großbauteilen: Ausführung, Bemessung und Lebensdauerbewertung*, Abschlussbericht DVS 12021 – IGF 18988, KIT Stahl- und Leichtbau, Versuchsanstalt für Stahl, Holz und Steine, Karlsruhe und Fraunhofer Institut für Werkstoffmechanik, Freiburg, 2019.

[103] Le Quilliec G., Lieurade H.-P., Bousseau M., Drissi-Habti M., Inglebert G., Macquet P., Jubin L.; *Fatigue Behaviour of Welded Joints Treated by High Frequency Hammer Peening: Part II, Numerical Study,* IIW Commission XIII, *Doc. XIII -2395-11*, 2011.

[104] Leblond J. B., Devaux J.; A New Kinetic Model for Anisothermal Metallurgical Transformations in Steels Including Effect of Austenite Grain Size, *Acta Metallurgica* **32,** pp. 137–146, 1984.

[105] Leblond J. B., Devaux J.; Mathematical Modelling of Transformation Plasticity in Steels I: Case of Ideal-Plastic Phases, *International Journal of Plasticity* **5,** pp. 551-572, 1989.

[106] Lee C. K., Chiew S. P., Jiang J.; 3D Residual Stress Modelling of Welded High Strength Steel Plate-to-Plate Joints, *Journal of Constructional Steel Research* **84**, pp. 94–104, 2013.

[107] Lee S. J., Lee Y. K.; Prediction of Austenite Grain Growth during Austenitization of Low Alloy Steels, *Materials & Design* **29**, pp. 1840–1844, 2008.

[108] Leitner M., Simunek D., Shah S. F., Stoschka M.; Numerical Fatigue Assessment of Welded and HFMI-Treated Joints by Notch Stress/Strain and Fracture Mechanical Approaches, *Advances in Engineering Software* **120**, pp. 96-106, 2018.

[109] Leitner M., Stoschka M., Eichlseder W.; Fatigue Enhancement of Thin-Walled, High-Strength Steel Joints by High-Frequency Mechanical Impact Treatment, *Welding in the World* **58** (1), pp. 29-39, 2013.

[110] Lim C. T., Stronge W. J.; Normal Elastic-Plastic Impact in Plane Strain, *Mathematical and Computer Modelling* **28** (4-8), pp. 323-340, 1998.

[111] Lindgren L.-E.; *Computational Welding Mechanics - Thermomechanical and Microstructural Simulations*, 1st Edition, Woodhead Publishing in Materials, Cambridge (UK), 2007.

[112] Loose T.; *Einfluß des transienten Schweißvorganges auf Verzug, Eigenspannungen und Stabilitätsverhalten axial gedrückter Kreiszylinderschalen aus Stahl*, Fakultät für Bauingenieur-, Geo- und Umweltwisseschaften, Universität Friedericiana zu Karlsruhe (TH) 2007.

[113] LS-DYNA, *Keyword User's Manual*, Livermore Software Technology Corporation (LSTC), Livermore California, 2016.

[114] LS-DYNA, *Theory Manual*, Livermore Software Technology Corporation (LSTC), Livermore California, 2016.

[115] Mangering I., Kroyer R., Koller M.; Experimental and Numerical Analysis of the Effectiveness of High-Frequency Peening Processes, *Steel Construction* **10** (1), pp. 54-66, 2017.

[116] Macedo M. Q., Cota A. B., Araújo F. G. d. S.; The Kinetics of Austenite Formation at High Heating Rates, *Rem Revista Escola de Minas* **64** (2), pp. 163–167, 2011.

[117] Magee A. C., Ladani L.; Temperature Dependency of Mechanical Behavior and Strain Rate Sensitivity of an Al-Mg Alloy with Bimodal Grain Size, *Materials Science and Engineering A* **582**, pp. 276-283, 2013.

[118] Marquis G. B., Barsoum Z.; *IIW Recommendations for the HFMI Treatment – For Improving the Fatigue Strength of Welded Joints*, 1st Edition, Springer Singapore (IIW Collection), Singapore, 2016.

[119] Marquis G. B., Mikkola E., Yildirim H. C.; Fatigue Strength Improvement of Steel Structures by High-Frequency Mechanical Impact: Proposed Fatigue Assessment Guidelines, *Welding in the World* **57**, pp. 803–822, 2013.

[120] Mataya M. C., Carr M. J.; The Bauschinger Effect in a Nitrogen-Strengthened Austenitic Stainless Steel, *Materials Science and Engineering* **57**, pp. 205-222, 1983.

[121] MATLAB and Statistics Toolbox Release R2015a, The MathWorks, Natick, Massachusetts, USA, 2015.

[122] May J., Höppel H. W., Göken M.; Strain Rate Sensitivity of Ultrafine-Grained Aluminium Processed by Severe Plastic Deformation, *Scripta Materialia* **53**, pp. 189-194, 2005.

[123] Mc Clung R. C., Sehitoglu H.; On the Finite Element Analysis of Fatigue Crack Closure – 1. Basic Modeling Issues, *Engineering Fracture Mechanics* **33** (2), pp. 237-252, 1989.

[124] Meguid S. A., Yang F., Zhuo C., Guagliano M.; 3D FE Analysis of Shot Peening of Strain rate Sensitive Material Account for Multiple and Repeated Impingement, *13th International Conference on Fracture, Beijing 16-21 June 2013*.

[125] Moradi R., McCoy M. L., Lankarani M.; *Impact Analysis of Mechanical Systems Using Stress Wave Propagation Methodology*, published in *Wave Propagation*, 1st Edition, Academy Publishing, pp. 209-247, 2014.

[126] Mullins J., Gunnars J.; *Influence of Hardening Model on Weld Residual Stress Distribution,* Research Report 2009:16, Inspecta Technology AB, Stockholm 2009.

[127] Nagel S., Knoedel P., Ummenhofer T.; Testing of Ultra-Low Cycle Fatigue at Complex Loading Scenarios, presented in the *2nd International Conference on Structural Integrity, Madeira 4-7 September 2017*, published in *Structural Integrity Procedia* **5**, pp. 1377-1384, 2017.

[128] Nart E., Celik Y.; A Practical Approach for Simulating Submerged Arc Welding Process Using FE Method, *Journal of Constructional Steel Research* **84**, pp. 62–71, 2013.

[129] Ngoula D. T., Beier H. T., Vormwald M.; Fatigue of Welded Joints Described by Elastic-Plastic Fracture Mechanics, *Stahlbau* **87** (5), pp. 491-496, 2018.

[130] Nitschke-Pagel T., Dilger K., Eslami H., Ummenhofer T.; Residual Stresses and Near Surface Material Condition of Welded High Strength Steels after Different Mechanical Post-Weld Treatments, *The Twentieth International Offshore and Polar Engineering Conference, Beijing 20-25 June 2010*.

[131] Norström L. Å.; The Influence of Nitrogen and Grain Size On Yield Strength in Type AISI 316L Austenitic Stainless Steel, *Metal Science* **11** (6), pp. 208-212, 1977.

[132] Oddy A., Goldak J., Mc Dill M.; Transformation Effects in the 3D Finite Element Analysis of Welds, *2nd International Conference in Welding Research*, Gatlinburg, 1989.

[133] Oliveira F. L. G., Andrade M. S., Cota A. B.; Kinetics of Austenite Formation During Continuous Heating in a Low Carbon Steel, *Materials Characterization* **58**, pp. 256–261, 2007.

[134] Ossenbrink R., Michailov V.; Thermomechanical Numerical Simulation with the Maximum Temperature Austenisation Cooling Time Model

(STAAZ), *Mathematical Modelling of Weld Phenoma* **8**, pp. 357–372, 2007.

[135] Paik J. K.; Cutting of a Longitudinally Stiffened Plate by a Wedge, *Journal of Ship Research* **38** (4), pp. 340–348, 1994.

[136] Perzyna P.; Fundamental Problems in Viscoplasticity, *Advances in Applied Mechanics* **9**, pp. 243-377, 1966.

[137] Ramberg W., Osgood W. R.; *Description of Stress-Strain Curves by three Parameters*, Technical Note No. 902, National Advisory Committee for Aeronautics, Washington DC, 1943.

[138] Rammerstorfer F. G., Fischer D. F., Mitter W., Bathe K. J., Snyder M. D.; On Thermo-Elastic-Plastic Analysis of Heat-Treatment Processes Including Creep and Phase Changes, *Computers and Structures* **13**, pp. 771-77, 1981.

[139] Radaj D.; *Heat Effects of Welding – Temperature Field, Residual Stress, Distortion*, 1st Edition, Springer Verlag, Berlin 1992.

[140] Rohr K.; *Evaluation of Weld-induced Residual Stresses in S690 Steel*, Diploma Thesis, KIT Faculty of Civil Engineering, Geo- and Environmental Sciences, Karlsruhe 2013.

[141] Sarkani S., Tritchkov V., Michaelov G.; An Efficient Approach for Computing Residual Stresses in Welded Joints, *Finite Elements in Analysis and Design* **35**, pp. 247-268, 2000.

[142] Schaumann P., Keindorf C.; Numerische Schweißsimulation gekoppelt mit einem anschließenden Hämmerprozess und integrierten lokalen Ermüdungsberechnungen, *Stahlbau* **79** (1), pp. 34-45, 2010.

[143] Schubnell J.; Experimentelle und numerische Untersuchung des Ermüdungsverhaltens von verfestigten Kerben und Schweißverbindungen nach dem Hochfrequenzhämmern, Dissertation, Fakultät für Maschinenbau, Karlsruhe, 2021.

[144] Schubnell J., Carl E., Farajian M., Gkatzogiannis S., Knödel P., Ummenhofer T., Wimpory R., Eslami H.; Residual Stress Relaxation in HFMI-Treated Fillet Welds After Single Overload Peaks, *IIW Commission XIII, Fatigue of Welded Components and Structures XIII-2829-19*, 2019.

[145] Schubnell J., Eichheimer C., Ernould C., Maciolek A., Rebelo-Kornmeier J., Farajian M.; The Influence of Coverage for High Frequency Mechanical Impact Treatment of Different Steel Grades, *Journal of Materials Processing Technology* **277**, 116437, 2020.

[146] Schubnell J., Gkatzogiannis S., Farajian M., Knoedel P., Luke T., Ummenhofer T.; *Rechnergestütztes Bewertungstool zum Nachweis der Lebensdauerverlängerung von mit dem Hochfrequenz-Hämmerverfahren (HFMI) behandelten Schweißverbindungen aus hochfesten Stählen*, Abschlussbericht DVS 09069 – IGF 19227 N, Fraunhofer Institut für Werkstoffmechanik, Freiburg und KIT Stahl- und Leichtbau, Versuchsanstalt für Stahl, Holz und Steine, Karlsruhe, 2020.

[147] Seyffarth P.; *Atlas SchweiB-ZTU-Schaubilder,* VEB Verlag Technik and DVS-Verlag, Berlin und Düsseldorf, 1982.

[148] Seyffarth P, Meyer B., Scharff A.; *Großer Atlas Schweiß-ZTU-Schaubilder*, DVS-Verlag, Düsseldorf, 1992.

[149] Shaw B. A., Korsunsky A. M., Evans J. T.; Surface Treatment and Residual Stress Effects on the Fatigue Strength of Carburised Gears, presented at the *12th European Conference on Fracture (ECF-12), Sheffield 14–18 September 1998.*

[150] Simunek D., Leitner M., Stoschka M.; *Numerical Simulation Loop to Investigate the Local Fatigue Behaviour of Welded and HFMI-Treated Joints*, IIW Document XIII-WG2-136-13.

[151] Smith M. C., Bouchard P. J., Turski M., Edwards L., Dennis R. J.; Accurate Prediction of Residual Stress in Stainless Steel Welds, *Computational Materials Science* **54**, pp. 312–328, 2012.

[152] Smith M. C., Smith A. C., Wimpory R., Ohms C.; A Review of the NeT Task Group 1 Residual Stress Measurement and Analysis Round Robin on a Single Weld Bead-on-Plate Specimen, *International Journal of Pressure Vessels and Piping* **120-121**, pp. 93-140, 2014.

[153] Statnikov E., Zhuravlev L. V., Alekseev A. F.; Ultrasonic Head for Strain Hardening and Relaxation Treatment (in Russian), USSR Inventor's Certificate No. 472782, *Bull Izobret* **21**, 1975.

[154] Stefani N., Bylya O., Reshetov A., Blackwell P.; On the Applicability of JMAK-Type Models in Predicting IN718 Microstructural Evolution, *Computer Methods in Materials Science* **17** (1), 59-68, 2017.

[155] Suresh S.; *Fatigue of Materials*, 2nd Edition, Cambridge University Press, Cambridge (UK), 2004.

[156] Symonds P. S., Chon C. T.; Approximation Techniques for Impulsive Loading of Structures of Time-Dependent Plastic Behaviour with Finite-

Deflections, *Mechanical Properties of Materials at High Strain Rates, Institute of Physics Conference Series* **21**, pp. 299–316, 1974.

[157] Symonds P. S.; *Survey of Methods of Analysis for Plastic Deformation of Structures under Dynamic Loading*, Department of the Navy Office of Naval Research Contract Nonr 3248(01)(X), Naval Ship Research and Development Centre, 1967.

[158] SYSWELD, SYSWELD Reference Manual, ESI Group, 2005.

[159] Tabor D.; *The Hardness of Metals*, Monographs on the Physics and Chemistry of Materials, Clarendon Press, Oxford, 1951.

[160] Tekriwal P., Mazumder J.; Transient and Residual Thermal Strain-Stress Analysis of GMAW, *Journal of Engineering Materials and Technology* **113** (3), pp. 336-343, 1991.

[161] Telljohann G., Dannemeyer S.; HiFIT – Hämmerverfahren zur Nachbehandlung geschweißter Stahlkonstruktionen, *Stahlbau* **77**, pp. 761-763, 2010.

[162] Telljohann G., Dannemeyer S.; HiFIT – Technische Entwicklung und Anwendung, *Stahlbau* **78** (9), pp. 622-626, 2009.

[163] Ting W., Dongpo W., Lixing H., Yufeng Z.; Discussion on Fatigue Design of Welded Joints Enhanced by Ultrasonic Peening Treatment (UPT), *International Journal of Fatigue* **31**, pp. 644-650, 2009.

[164] Ueda Y., Kim Y. C., Chen C., Tang Y. M.; Mathematical Treatment of Phase Transformation and Analytical Calculation Method of Restrain Stress-Strain, *Transactions of Japan Welding Institute* **14** (1), pp. 153-162, 1985.

[165] Ueda Y., Yamakawa T.; Thermal Stress Analysis of Metals with Temperature Dependent Mechanical Properties, *Transactions of the Japan Welding Research Institute* **2** (2), pp. 90-100, 1971.

[166] Ullah H., Ullah B., Muhammad R.; Explicit Dynamics Simulation of Shot Peening Process Induced by Various Types of Shot Impacts, *Journal of the Brazilian Society of Mechanical Sciences and Engineering* **40** (110), 2018.

[167] Ummenhofer T.; *REFRESH: Lebensdauerverlängerung bestehender und neuer geschweißter Stahlkonstruktionen*, Abschlussbericht D 761, KIT Stahl- und Leichtbau, Versuchsanstalt für Stahl, Holz und Steine, Karlsruhe, 2009.

[168] Ummenhofer T., Engelhardt I., Knoedel P., Gkatzogiannis S., Weinert J., Loeschner D.; *Erhöhung der Ermüdungsfestigkeit von Offshore-Windenergieanlagen durch Schweißnahtnachbehandlung unter Berücksichtigung des Korrosionseinflusses*, Schlussbericht, DVS 09069 – IGF 18457 N, KIT Stahl- und Leichtbau, Versuchsanstalt für Stahl, Holz und Steine, Karlsruhe und Hochschule für angewandte Wissenschaften München, Labor für Stahl- und Leichtmetallbau, 2018.

[169] Ummenhofer T., Herion S., Puthli R., Weidner P.; Effects of Post Weld Treatment and Loading Condition on the Fatigue Life of Ultra High Strength Steels, *21st International Offshore and Polar Engineering Conference, Maui Hawaii 19-24 June 2011.*

[170] Ummenhofer T., Knödel P., Rohr K.; *Stahlwerkstoffe, Schweißtechnik und Ermüdung Teilmodul Schweißtechnik*, Skript, Lehrstuhl für Stahl und Leichtmetallbau, Karlsruhe 2017.

[171] Ummenhofer T., Weich I., Nitschke-Pagel T.; Lebens- und Restlebensdauerverlängerung geschweißter Windenergieanlagentürme und anderer Stahlkonstruktionen durch Schweißnahtnachbehandlung, *Stahlbau* **74** (6), pp. 412-422, 2005.

[172] Ummenhofer T., Weidner P.; Improvement Factors for the Design of Welded Joints subjected to High Frequency Mechanical Impact Treatment, *Steel Construction* **6** (3), pp. 191-199, 2013.

[173] Ummenhofer T., Weidner P., Zinke T.; New and Existing Bridge Constrictions – Increase of Fatigue Strength of Welded Joints by High Frequency Mechanical Impact Treatment, *International Scientific Conference Road Research and Administration, "CAR 2013", Bucharest 4-5 July 2013.*

[174] Weich I., Ummenhofer T., Nitschke-Pagel T., Dilger K., Eslami H.; Fatigue Behaviour of Welded High-Strength Steels after High Frequency Mechanical Post-Weld Treatments, *Welding in the World* **53** (11-12), pp. 322-332, 2009.

[175] Weidner P., Herion S., Puthli R., Ummenhofer T.; Lebensdauerverlängerung von geschweißten Verbindungen aus Rundhohlprofilen. *Stahlbau* **80** (7), pp. 502-512, 2011.

[176] Weidner P., Mehdianpour M., Ummenhofer T.; Ermüdungsfestigkeit einseitig geschweißter Stumpfnähte von Kreishohlprofilen, *Stahlbau* **85** (9), pp. 620-629, 2016.

[177] Wohlfahrt H.; Die Bedeutung der Austenitumwandlung für die Eigenspannungsentstehung beim Schweißen, *Härterei - technische Mitteilungen* **41** (5), pp. 248-257, 1986.

[178] Wohlfahrt H., Nitschke-Pagel T., Dilger K., Siegele D., Brand M., Sakkiettibutra J., Loose T.; *Residual Stress Calculations and Measurements – Review and Assessment of the IIW Round Robin Results*, IIW Document 2224:120-140, 2012.

[179] Yamazaki K., Yamamoto E., Suzuki K., Koshiishi F., Tashiro S., Tanaka M., Nakata K.; Measurement of Surface Temperature of Weld Pools by Infrared Two Colour Pyrometry, *Science and Technology of Welding and Joining* **15**, pp. 40-47, 2010.

[180] Yan W., Sun Q., Hodgson P. D.; Determination of Plastic Yield Stress from Spherical Indentation Slope Curve, *Materials Letters* **62**, pp. 2260 – 2262, 2008.

[181] Yildirim H. C., Marquis G. B.; A Round Robin Study of High-Frequency Mechanical Impact (HFMI) - Treated Welded Joints subjected to Variable Amplitude Loading, *Welding in the World* **57**, pp. 437–447, 2013.

[182] Yildirim H. C., Marquis G. B.; Fatigue Strength Improvement Factors For High Strength Steel Welded Joints Treated By High Frequency Mechanical Impact, *International Journal of Fatigue* **44**, pp. 168-176, 2012.

[183] Yildirim H. C., Marquis G. B.; Overview of Fatigue Data for High Frequency Mechanical Impact Treated Welded Joints, *Weld World* **56** (7-8), pp. 82-96, 2013.

[184] Wilson E. A., Medina S. F.; Application of Koistinen and Marburger's Athermal Equation for Volume Fraction of Martensite to Diffusional Transformations Obtained on Continuous Cooling 0.13 %C High Strength Low Alloy Steel, *Materials Science and Technology* **16** (6), pp. 630-633, 2013.

[185] Yildirim H. C., Marquis G. B., Barsoum Z.; Fatigue Assessment of High Frequency Mechanical Impact (HFMI) – Improved Fillet Welds by Local Approaches, *International Journal of Fatigue* **52**, pp. 57-67, 2013.

[186] Yuan K. L., Sumi Y.; *Welding Residual Stress and its Effect on Fatigue Crack Propagation after Overloading*, published in Guedes Soares C., Romanoff J.; *Analysis and Design of Marine Structures,* 1st Edition, Taylor & Francis Group, London, 2013.

[187] Yuan K. L., Sumi Y.; Modelling of Ultrasonic Impact Treatment (UIT) of Welded Joints and its Effect on Fatigue Strength, *Frattura ed Integrità Strutturale* **34**, pp. 476–486, 2015.

[188] Zhang Z., Farrar R.; *An Atlas of Continuous Cooling Transformation (CCT) Diagrams Applicable to Low Carbon Low Alloy Weld Metals*, 1st Edition, Bourne Press Ltd, Bournemouth 1995.

[189] Zheng J., Ince A., Tang L.; Modelling and Simulation of Weld Residual Stresses and Ultrasonic Impact Treatment of Welded Joints, *7th International Conference on Fatigue Design, Senlis 29-30 November 2017, Procedia Engineering* **213**, pp. 36–47, 2018.

[190] Zhu X. K., Chao Y. J.; Effects of Temperature-Dependent Material Properties on Welding Simulation, *Computers and Structures* **80**, pp. 967-976, 2002.

Appendix A – Investigated Materials

Inspection Certificates

<u>S355 from project *HFH-Simulation*</u>

geprüft 26.04.2017 gei

thyssenkrupp Steel Europe — thyssenkrupp

Werks-Nr. / Works-No. / No de l'usine A08	Zeugnis-Nr. / Certificate-No. / No de certificat A03	Sendungs-Nr. / Shipment-No. / No de envoi	Seite-Nr. / Page-No. / Page-No.
4243698	1810502001	46744706	1

thyssenkrupp Steel Europe AG · 47161 Duisburg A01

A06

LIEBHERR WERK BIBERACH GMBH

MEMMINGER STR. 120
D 88400 BIBERACH

BESCHEINIGUNG ÜBER MATERIALPRÜFUNGEN EN 10204
DOCUMENT ON MATERIAL TESTS EN 10204
DOCUMENT DE CONTROL DES MATERIAUX EN 10204

A02
ABNAHMEPRUEFZEUGNIS 3.1
INSPECTION CERTIFICATE 3.1

Bestell-Nr. / Order-No. / No de commande 013446

A07.1-A07.5

31.03.2017
0203 52 75220
0203 52 75213
abnahme_zeugnisse.grobblech@thyssenkrupp.com A05

Werkstoff ; Quality ; Matériau / Lieferbedingungen ; Specification ; Conditions de livraison B02/B03
S355J2+N EN10025-2 04.05
TOL.EN10029 KL.A/N OB EN 10163 KL.B/3

Kennzeichnung: / Marking: / Marque: B06 WERKSTOFF; SCHMELZ-NR.; FERTIGUNGS-/PROBE-NR.
MATERIAL, HEAT-NO., MANUFACTURING/SAMPLE-NO.

Zeichen des Lieferwerkes: / Supplier's mark: / Marque d'usine: A04

B01
ERZEUGNISFORM
TYPE OF PRODUCT

GROBBLECH, BESAEUMTE KANTEN
HEAVY PLATES, TRIMMED EDGES

LISTE DER MATERIALIDENTEN
LIST OF MATERIAL IDENTS

POS. / ITEM	B07 PAKET / BUNDLE	B07 BLECH-NR. / PLATE-NO.	B07 SCHMELZ-NR. / HEAT-NO.	B08 STUECK ZAHL / NUMBER PIECES	B13 GEWICHT GEWOG. / WEIGHT
	B09 x B10 x B11				Kg
002	10,0 x 2500,0 x 6000 [mm]				
		76478101	622050	1	1.184
		76478102	622050	1	1.184
		76478203	622050	1	1.184
		76478204	622050	1	1.184
		76478305	622050	1	1.184
		76478306	622050	1	1.184
		76484101	622050	1	1.184
		76484102	622050	1	1.184
		76484203	622050	1	1.184
		76484204	622050	1	1.184
		76484305	622050	1	1.184
		76484306	622050	1	1.184
		74838102	670800	1	1.184
		74838203	670800	1	1.184
		74838204	670800	1	1.184
		74838305	670800	1	1.184
		74838306	670800	1	1.184
		76482101	670800	1	1.184
		76482102	670800	1	1.184
			*	19	22.496
			**	19	22.496

thyssenkrupp Steel Europe AG

Abnahme -FOLGESEITE-

Werks-Nr. / Works-No. / No de l'usine A08	Zeugnis-Nr. / Certificate-No. / No de certificat A03	Sendungs-Nr. / Shipment-No. / No de envoi	Seite-Nr. / Page-No. / Page-No.
4243698	1810502001	46744706	2

TRANSPORT-NR.
TRANSPORT-NO.
MR-P 620

CHEMISCHE ZUSAMMENSETZUNG DER SCHMELZE IN % C71-C99
CHEMICAL COMPOSITION OF THE LADLE SAMPLES %

B07

SCHMELZ-NR	C	SI	MN	P	S	AL-G	B-G	CR	CU	MO	N
622050	,156	,187	1,400	,013	,0019	,027	,0002	,028	,014	,003	,0075
670800	,156	,165	1,420	,012	,0037	,026	,0001	,033	,012	,008	,0059

	NB	NI	TI	V	CE1
622050	,015	,029	,014	,002	,40
670800	,015	,026	,014	,003	,40

CE1 = C+MN/6+(CR+MO+V)/5+(NI+CU)/15

C70 SCHMELZVERFAHREN OXYGENSTAHL
C70 HEAT PROCESS OXYGEN STEEL

MECHANISCHE EIGENSCHAFTEN Z U G V E R S U C H
MECHANICAL CHARACTERISTICS T E N S I L E T E S T

B07 SCHM.-NR.	C00 PROBE-NR.	C01/02 LAGE	B05 ZUST.	C10 FORM	ALTER	C03 TEMP. °C	C11 R MPa	R Art	C12 Rm MPa	R/Rm %	LO mm	C13 A %	Rm*A
622050	*76363	0401	0004	0002	0006	+20	420	RE H	545	77	105	29	15805
622050	76478	0401	0004	0002	0006	+20	427	RE H	551	77	100	28	15428
670800	*75880	0401	0004	0002	0006	+20	411	RE H	538	76	120	25	13450
670800	*75891	0401	0004	0002	0006	+20	401	RE H	539	74	120	30	16170

MECHANISCHE EIGENSCHAFTEN K E R B S C H L A G B I E G E V E R S U C H
MECHANICAL CHARACTERISTICS I M P A C T T E S T

B07 SCHM.-NR.	C00 PROBE-NR.	C01/02 LAGE	B05 ZUST.	C40 FORM	C41 B mm	ALTER	C03 PRUEF-TEMP. °C	C42 ARBEIT [Joule] 1	C42 2	C42 3	C43 M
622050	*76463	0101	0004	0007	10,00	0006	-20	125	147	115	129
622050	*76476	0101	0004	0007	10,00	0006	-20	148	156	156	153
670800	*75916	0101	0004	0007	10,00	0006	-20	230	224	203	219

* PROBEBLECH NICHT IN LIEFERUNG ENTHALTEN
* SAMPLE PLATE NOT INCLUDED IN DELIVERY

L E G E N D E N
L E G E N D S

ALTER / AGED — ART / TYPE

PROBENZUSTAND / STAT.
0004 = NORMALISIERT / NORMALIZED

ALTERUNG / AGED
0006 = UNGEALTERT / NOT AGED

PROBENFORM KERBSCHLAG / TYPE IMPACT TEST
0007 = CHARPY- V / CHARPY- V

PROBENLAGE (IST) / POSIT (IST)
0101 = LAENGS KOPF OBERFLAECHE / LONG. TOP S
0401 = QUER KOPF OBERFLAECHE / TRANS. TOP S.

PROBENFORM ZUGVERSUCH / TYPE TENSILE TEST
0002 = FLACHZUG / FLAT TENSILE TEST

L I E F E R Z U S T A N D P R O D U K T
S T A T U S P R O D U C T

002 NORMALISIERT

thyssenkrupp Steel Europe AG

Abnahme -FOLGESEITE-

Werks-Nr. / Works-No. / No de l'usine	Zeugnis-Nr. / Certificate-No. / No de certificat	Sendungs-Nr. / Shipment-No. / No de envoi	Seite-Nr. / Page-No. / Page-No.
A08 4243698	A03 1810502001	46744706	3

NORMALIZED

ERGEBNIS DER BESICHTIGUNG UND MASSPRUEFUNG: KEINE BEANSTANDUNG

RESULT OF SURFACE CONTROL AND DIMENSIONAL CHECK: SATISFACTORY

Z04

CE

0769

thyssenkrupp Steel Europe AG, 47161 Duisburg

17

0769-CPR-VAS-00439-1

EN 10025-1

Warmgewalzte Baustahlprodukte

Verwendungen: Hochbauten und Ingenieurbauwerke

Grenzabmaße und Formtoleranzen:

Grobblech EN10029 KL. A/ EN10029 KL. N

Dehnung	
Zugfestigkeit	
Streckgrenze	S355J2 / 1.0577
Kerbschlagarbeit	EN 10025-2:2004 *)
Schweißeignung	

Dauerhaftigkeit: Keine Leistung festgestellt

Regulierter Stoff: Keine Leistung festgestellt

*) DoP für Stahl S355J2 / 1.0577 / EN 10025-2:2004 *) siehe Dokument Nr. GBQ-0017-S355J2-TKSE-CPR-29042013 auf TKSE-Homepage http://www.thyssenkrupp-steel-europe.com/de/publikationen/produktinformationen/bauproduktenverordnung.jsp

*) DoP for Steel S355J2 / 1.0577 / EN 10025-2:2004 *) see Document No. GBQ-0017-S355J2-TKSE-CPR-29042013 on TKSE-Homepage http://www.thyssenkrupp-steel-europe.com/de/publikationen/produktinformationen/bauproduktenverordnung.jsp

Z02
ABNAHMEBEAUFTRAGTER DES HERSTELLERS
THE MANUFACTURER'S AUTHORIZED INSPECTION REPRESENTATIVE
LE REPRÉSENTANT AUTORISÉ DU CONTRÔLEUR DU PRODUCTEUR

Kern

Z03 Prof. Dr.-Ing. Kern

thyssenkrupp Steel Europe AG

Abnahme

Es wird bestätigt, dass die Lieferung den Vereinbarungen bei der Bestellung entspricht.
We hereby certify that the above mentioned materials have been delivered in accordance with the terms of order.
Nous attestons que les produits livrés sont conformes aux stipulations de la commande.
Z01

S690 from project *HFH-Simulation*

TATA

TATA STEEL UK LIMITED A1
Park Street
Motherwell ML1 1PU
Tel: +44 (0)1698 266100
Fax: +44 (0)1698 266100 Ext. 227

ABNAHMEPRUEFZEUGNIS A2

In Accordance with EN10204 3.1
Material hergestellt gemass einem ISO 9001 zertifizierten Management System

PAGE 1

Datum Z1	8/ 6/12
Zeugnis Nr. A3	DALZ/12/01181195/1
	PLATE REF 5915155
Tata Steel Kennzeichen A8	666090
Werks Nr. A8	261075/001
Kundenbestell Nr. A7	E1158329/1 Q&T — Contract No.:- 2/041
Werkstoff/Lieferbedingungen B1/B2	RQT 701/EN10025-6:2004 S690QL
	STEEL PLATE 12.000m X 2.500m X 10.00mm Piece Weight 2.355Tonnes Quenched at +930C for 20 mins. Tempered at +600C for 30 mins.
Abnahmegesellschaft A11	QWN
Besteller A6	

CE 0038

ORDER MARKS :- PPU68609/B....

L=Langs T=Quer Z=Senkrecht S=Oberflache M=Dickenmitte Q=1/4 Blechdicke TH=1/6 Blechdicke BS=Unterseite A=Supply Condition N=Normalisiert SR=Spannungsarmgegluht SA=Geallert

B10 Schmelz Nr. B6	Blech Nr. B7	Probe Nr. C00	C2 Richtung	Proben-zustand B5	Streckgrenze Re C11 (MPa)	Zugfestigkeit Rm C12 (MPa)	Re/Rm % C14	Dehnung % A Lo=5.65/√So C13	C2 Richtung	C1 Position	Proben-zustand B5	Charpy-Kerbschlagarbeit 10x7.5x2mmV C40/C41, Einheiten C42: Joules				C43 Av.
	Sollwerte Min./Max.															
68/4960	7007E155		T	A	774	828		20	L	S	A	-40C	139	145	110	131
									T	S	A	-40C	62	37	40	46

Erschmelzung C70: B.O.S. — Continuously cast slab.
Sonstige Proben C50-C69

ANALYSE % C71-C92		C	Si	Mn	P	S	Cr	Mo	Ni	Tot Al	Cu	Nb	B	Sn	Ti	N	V	Zr	CEV
Sollwerte	Min./Max.																		
Cast/Heat		.14	.40	1.38	.007	.002	.01	.002	.02	.030	.02	.030	.002	.002	.024	.005	.061	TR	.39
Sollwerte	Min./Max.																		

Marking Certificate Number = 0038/CPD/20050001A

In Namen von Tata Steel UK Limited.
Die Ergebnisse werden von Tata Steel UK Limited bescheinigt und entsprechen den Anforderungen gemass Produktbeschreibung. Z1

R.J.Preston. Chef der Abnahme, Scunthorpe

S960 from project *HFH-Simulation*

STAHLHANDEL · SCHNEIDBETRIEB · MASCHINEN- UND ANLAGENBAU

STAHLSERVICE DILLINGEN

pink & wagner GMBH

PINK & WAGNER GMBH · POSTFACH 1629 · 66749 DILLINGEN

Fraunhofer-Institut
für Werkstoffmechanik IWM
Wöhlerstraße 11
79108 Freiburg

Gewerbegebiet Nord
Dieselstraße 40-48
D-66763 Dillingen/Saar
Telefon 00 49 (0) 68 31 76879-0
Telefax 00 49 (0) 68 31 76879-68
www.pinkundwagner.de

Ihr Zeichen | Ihre Nachricht | Unser Zeichen | Datum

Umstempelungsbescheinigung Nr.: 14212

Ihre Bestellung: 019/243432/921

Unser Auftrag : 223750

Wir bestätigen Ihnen hiermit, dass die von uns gelieferten Erzeugnisse:

1 Stk. 500 x 1500 x 10 mm

aus der Abmessung: 10 x 2500 x 12000 mm

ausgewiesen durch das Werksabnahmeprüfzeugnis Nr. 16031687 nach EN 10204 / 3.1 des Werkes SSAB geschnitten wurden und die Stempelung vor dem Trennen übertragen worden ist.

Stempelung wie folgt:

Erschmelzungsverfahren:	
Werkstoff	: STRENX 960 E
Schmelze-Nr.	: 118232
Probe-Nr.	: 713515

Zum Zeichen der erfolgten Prüfung und der ordnungsgemäßen Umstempelung wurden die Erzeugnisse mit folgendem Stempel versehen. PW 8

Das Umstempeln erfolgt mit Zustimmung des TÜV Saarland
(Vereinbarung vom 29.09.2016 Prüf-Nr.: 71 201 16 U 0182)

Dillingen, den 14. August 2017 Unterschrift: ________

Geschäftsführung: Hans-Georg Pink, Vero Caspar, Susanne Pink
Amtsgericht Saarbrücken, HRB 24394
Gerichtsstand für beide Teile ist Saarlouis

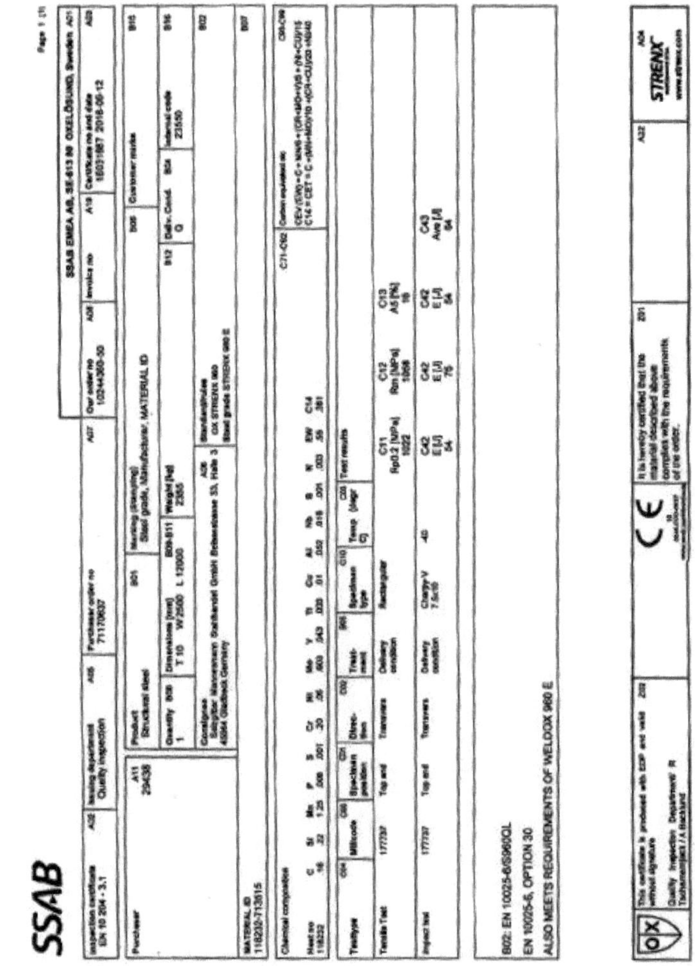
SSAB

Page 1 (1)

SSAB EMEA AB, SE-613 80 OXELÖSUND, Sweden A01

Inspection certificate A02	Issuing department A05	Purchaser order no A07	Our order no A08	Invoice no A19	Certificate no and date A03
EN 10 204 - 3.1	Quality inspection	71170637	10244360-50		18031887 2018-08-12

Purchaser A11: 29438

Product B01: Structural steel

Marking (Stamping) B05: Steel grade, Manufacturer, MATERIAL ID

Customer marks B15

Quantity B08: 1

Dimensions [mm] B09-B11: T 10 W2500 L 12000

Weight [kg] B12: 2355

Deliv. Cond. B04: Q

Internal code B16: 23550

Consignee A06: Salzgitter Mannesmann Stahlhandel GmbH Bebelstrasse 33, Halle 3 45964 Gladbeck Germany

Standard/rules B02: OX STRENX 960; Steel grade STRENX 960 E

MATERIAL ID B07: 118232-713615

Chemical composition C71-C92

Heat no	C	Si	Mn	P	S	Cr	Ni	Mo	V	Ti	Cu	Al	Nb	B	N	EW	C14
118232	.16	.22	1.25	.008	.001	.20	.06	.600	.043	.003	.01	.052	.018	.001	.003	.55	.381

Carbon equivalent etc C93-C99

CEV (EW) = C + MN/6 + (CR+MO+V)/5 + (NI+CU)/15

C14 = CET = C + (MN+MO)/10 + (CR+CU)/20 + NI/40

Testtype C04	Millcode C05	Specimen position C01	Direction C02	Treatment B05	Specimen type C10	Temp (degr C) C03	Test results			
Tensile Test	177737	Top end	Transvers	Delivery condition	Rectangular		C11 Rp0.2 [MPa] 1022	C12 Rm [MPa] 1058	C13 A5 [%] 16	
Impact test	177737	Top end	Transvers	Delivery condition	Charpy-V 7.5x10	-40	C42 E [J] 54	C42 E [J] 75	C42 E [J] 64	C43 Ave [J] 64

B02: EN 10025-6/S960QL

EN 10025-6, OPTION 30

ALSO MEETS REQUIREMENTS OF WELDOX 960 E

This certificate is produced with EDP and valid without signature Z02

Quality Inspection Department/ R Tschamernjack / A Backlund

It is hereby certified that the material described above complies with the requirements of the order. Z01

A22

A04 STRENX www.strenx.com

Welding Procedure Specifications

S355 from project *HFH-Simulation*

LIEBHERR Werk Biberach GmbH

Berechnung Streckenenergie

Schweißaufgabe: Instandsetzung Großbauteile

Werkstoff: S355J2+N an S355J2+N

Verfahren: MAG 135 | **Roboter:** Clsso R350 | **Bediener:** Hr. Weißbrodt

Probe Nr.: REPA 1-4 | **Gerät:** Cloos | **Schweißer::** -

Gasdurchfluß: 19,0 l / min | Quinto GLC 603 | **Schweißaufsicht:** Hr. Schmid

Draht-Ø: 1,2 mm | **Drahtanzahl:** 1 | **SFI:** Hr. Hänsler

Schw.-draht: G4Si1 (M21) | **Pulver:** -- | **WPS-Nr.:** --

Vorwärtemp.: RT | **Brennerhaltung:** neutral (3°stechend) | **Probe-/ Bauteil:** Mutterkreuzstoß

Raupe Nr.	Probe	Schweiß-verfahren	Spannung	Stromstärke	Nahtlänge	Schweißzeit	Streckenenergie	thermischer Wirkungsgrad	**Wärmeeinbringung**	Schweißge-schwindigkeit	Drahtvor-schub
[-]		[-]	[V]	[A]	[cm]	[sec]	[kJ]	[-]	**[kJ/mm]**	[cm/min]	[m/min]
1	REPA 1	135	29,6	246	123,0	185,0	1,10	0,8	**0,88**	39,9	8,5
2	REPA 1	135	29,4	249	123,0	185,0	1,10	0,8	**0,88**	39,9	8,5
3	REPA 1	135	29,5	248	123,0	185,0	1,10	0,8	**0,88**	39,9	8,5
4	REPA 1	135	29,5	248	123,0	185,0	1,10	0,8	**0,88**	39,9	8,5
1	REPA 2	135	29,4	249	123,0	185,0	1,10	0,8	**0,88**	39,9	8,5
2	REPA 2	135	29,4	248	123,0	185,0	1,10	0,8	**0,88**	39,9	8,5
3	REPA 2	135	29,3	247	123,0	185,0	1,09	0,8	**0,87**	39,9	8,5
4	REPA 2	135	29,4	247	123,0	185,0	1,09	0,8	**0,87**	39,9	8,5
1	REPA 3	135	29,4	247	123,0	185,0	1,09	0,8	**0,87**	39,9	8,5
2	REPA 3	135	29,4	247	123,0	185,0	1,09	0,8	**0,87**	39,9	8,5
3	REPA 3	135	29,5	248	123,0	185,0	1,10	0,8	**0,88**	39,9	8,5
4	REPA 3	135	29,4	248	123,0	185,0	1,10	0,8	**0,88**	39,9	8,5
1	REPA 4	135	29,4	247	123,0	185,0	1,09	0,8	**0,87**	39,9	8,5
2	REPA 4	135	29,4	248	123,0	185,0	1,10	0,8	**0,88**	39,9	8,5
3	REPA 4	135	29,4	247	123,0	185,0	1,09	0,8	**0,87**	39,9	8,5
4	REPA 4	135	29,5	247	123,0	185,0	1,10	0,8	**0,88**	39,9	8,5

Kreuzstöße mit Thermoelementen

1	1	135	29,4	248	15,0	23,0	1,12	0,8	**0,89**	39,1	8,5
2	1	135	29,3	248	15,0	23,0	1,11	0,8	**0,89**	39,1	8,5
1	2	135	29,4	249	15,0	23,0	1,12	0,8	**0,90**	39,1	8,5
2	2	135	29,3	249	15,0	23,0	1,12	0,8	**0,89**	39,1	8,5
1	3	135	29,4	248	15,0	23,0	1,12	0,8	**0,89**	39,1	8,5
2	3	135	29,4	249	15,0	23,0	1,12	0,8	**0,90**	39,1	8,5

S690 – S960 from project *HFH-Simulation*

	WPS	WPS-Nr: Rev
		MB_10.10_960_A1

Ort:	Wegscheid	Prüfer oder Prüfstelle:	
WPQR-Nr.:		Art der Vorbereitung und Reinigung:	mechan. Bearbeitung
Schweißerqualifikation:		Bearbeitung der Wurzellage:	Schleifen
Schweißprozess:	135-(MAG)	Spezifikation Grundwerkstoff(e):	Gruppennr. ISO 15608:
Nahtart:	Kehlnaht	1) [1.8925], S960QL	3.2
Kunde:	Frauenhofer Institut	2) [1.8925], S960QL	3.2
Auftrags-Nr.:		Werkstoffdicke:	10 mm
Zeichnungs-Nr.:		Außendurchmesser:	
Teile-Nr.:		Schweißposition:	PB

Maße:

t = 10 mm
a = 4 mm

Gestaltung der Verbindung

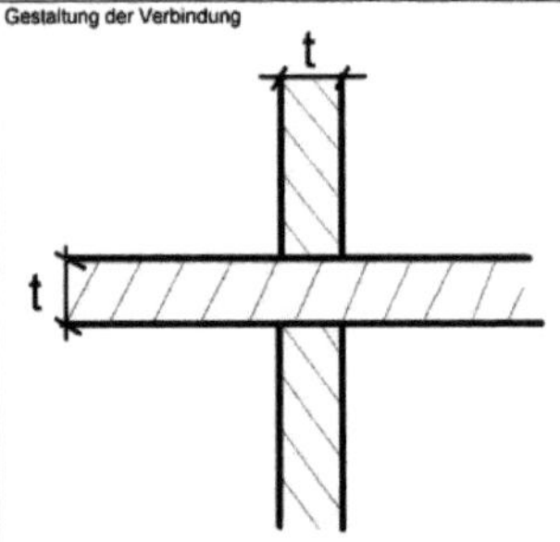

Schweißfolge

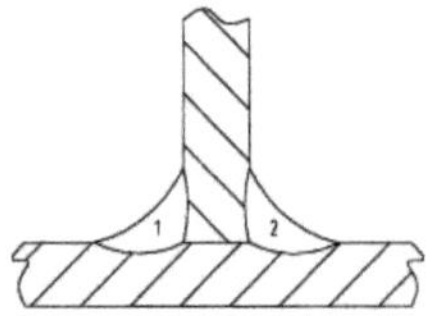

Bemerkung:

Einzelheiten für das Schweißen

	Schweißlage	Prozess	Ø Schweiß-zusatz [mm]	Strom	Spannung [V]	Stromart / Polung	Draht-vorschub-geschw.	Schweiß-geschwindig-keit [cm/min]	Wärme-einbringung [kJ/mm]
1)	Wurzellage	135	1,0	155 - 175 A	24,5-26,5	= / +	9-10 m/min	38-40	0,456 - 0,586

Schweißzusatz / Schweißpulver — Sondervorschriften für Trocknung

	Bezeichnung	Markenname	Hersteller	Zeit [h]	Temperatur [°C]
1)	DIN EN 757 - E 89 4 Mn2Ni1CrMo B 4 2 H5	Union X90	Böhler		

Schutzgas

	Typ	Markenname	Hersteller	Durchfluss [l/min]	Vorström- zeit [s]	Nachström-zeit [s]
1)	Schweißen: M21-ArC-18			15-18		

Weitere Informationen

	Parameter / Wert

Pendeln: Strichraupe
Vorwärmtemperatur[°C]:
Zwischenlagentemperatur [°C]:

Datum / Erstellt: 11.04.2017 Franz-Josef Gottinger	Datum / Geprüft:	Datum / Freigegeben:
Unterschrift	Unterschrift	Unterschrift

Stahlbau Wegscheid GmbH, Kasberger Str. 25, 94110 Wegscheid
Erstellt mit schweissassistent® von hsk-welding solutions

	WPS	WPS-Nr: MB_8.12_PB_S960_FW	Rev.

Ort:		Prüfer oder Prüfstelle:		
WPQR-Nr.:		Art der Vorbereitung und Reinigung:		
Schweißerqualifikation:		Bearbeitung der Wurzellage:		
Schweißprozess:	135-(MAG)	Spezifikation Grundwerkstoff(e):		Gruppennr. ISO 15608:
Nahtart:	Kehlnaht	1) S960QL		3.2
Kunde:	Frauenhofer Institut	2) S960QL		3.2
Auftrags-Nr.:		Werkstoffdicke:	8 mm	
Zeichnungs-Nr.:		Außendurchmesser:		
Teile-Nr.:		Schweißposition:	PB	

Maße:

t1 = 8 mm
t2 = 12 mm
a = 5 mm

Gestaltung der Verbindung

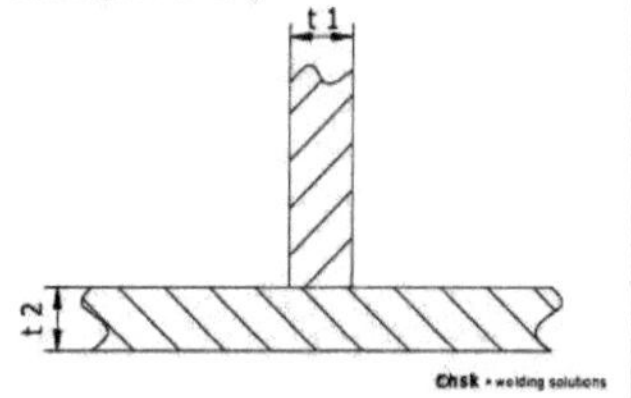

Schweißfolge

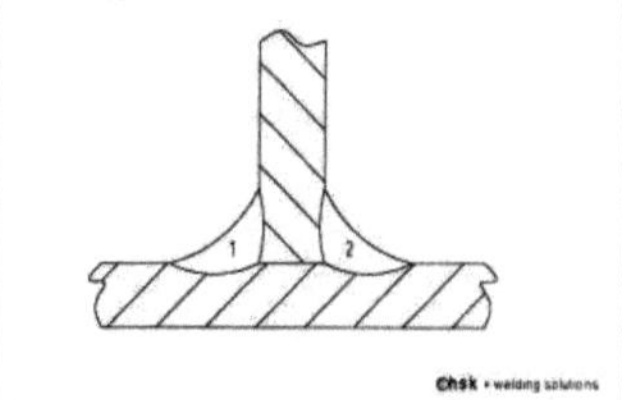

Bemerkung:
Lage 1 und 2 wurden parallel geschweißt.

Einzelheiten für das Schweißen

	Schweißlage	Prozess	Ø Schweiß-zusatz [mm]	Strom	Spannung [V]	Stromart / Polung	Draht-vorschub-geschw.	Schweiß-geschwindig-keit [cm/min]	Wärme-einbringung [kJ/mm]
1-2)	Wurzellage	135	1,0	185-205 A	28,5 - 30,5	= / +	12-13 m/min	31-33	0,767 - 0,968

Schweißzusatz / Schweißpulver — Sondervorschriften für Trocknung

	Bezeichnung	Markenname	Hersteller	Zeit [h]	Temperatur [°C]
1-2)	DIN EN 757 - E 89 4 Mn2Ni1CrMo B 4 2 H5	Union X90	Böhler		

Schutzgas

	Typ	Markenname	Hersteller	Durchfluss [l/min]	Vorström- zeit [s]	Nachström-zeit [s]
1-2)	Schweißen: M21-ArC-18			15-18		

Weitere Informationen

	Parameter / Wert

Pendeln: Strichraupe
Vorwärmtemperatur[°C]:
Zwischenlagentemperatur [°C]:

Datum / Erstellt:	Datum / Geprüft:	Datum / Freigegeben:
11.04.2017 Franz-Josef Gottinger		
Unterschrift	Unterschrift	Unterschrift

Stahlbau Wegscheid GmbH, Kasberger Str. 25, 94110 Wegscheid
Erstellt mit schweissassistent® von hsk-welding solutions

CMM

HT36

CMM 2 (T_{max} = 950 °C, t_{85} = 10 s)												
Temp [°C]	microstructure					σ_{yi} [MPa]					σ_y [MPa]	a^{se} [-]
	PM	AY	FE ,PE	BA	MA	AY	FE, PE	BA	MA	PM		
950	63 %	37 %	0 %	0 %	0 %	53	38	38	38	38	43	18
520	63 %	37 %	0 %	0 %	0 %	114	177	363	483	175	152	18
320	63 %	7 %	0 %	30 %	0 %	148	206	445	569	270	313	15
270	63 %	0 %	0 %	30 %	7 %	111	250	485	612	188	307	15
20	63 %	0 %	0 %	30 %	7 %	200	258	490	620	350	411	15

CMM 3 (T_{max} = 1150 °C, t_{85} = 10 s)												
Temp [°C]	microstructure					σ_{yi} [MPa]					σ_y [MPa]	a^{se} [-]
	PM	AY	FE ,PE	BA	MA	AY	FE, PE	BA	MA	PM		
1150	31 %	69 %	0 %	0 %	0 %	29	14	14	14	15	25	21
520	31 %	69 %	0 %	0 %	0 %	114	177	363	483	175	133	21
320	31 %	14 %	0 %	55 %	0 %	148	206	445	569	270	350	16
270	31 %	0 %	0 %	55 %	14 %	111	250	485	612	188	410	14
20	31 %	0 %	0 %	55 %	14 %	200	258	490	620	350	465	14

CMM 4 (T_{max} = 1350 °C, t_{85} = 10 s)												
Temp [°C]	microstructure					σ_{yi} [MPa]					σ_y [MPa]	a^{se} [-]
	PM	AY	FE ,PE	BA	MA	AY	FE, PE	BA	MA	PM		
1350	0	100 %	0 %	0 %	0 %	5	5	5	5	5	5	24
520	0	100 %	0 %	0 %	0 %	114	177	363	483	175	114	24
320	0	20 %	0 %	80 %	0 %	148	206	445	569	270	385	16
270	0	0 %	0 %	80 %	20 %	111	250	485	612	188	510	14
20	0	0 %	0 %	80 %	20 %	200	258	490	620	350	516	14

CMM 5 (T_{max} = 950 °C, t_{85} = 35 s)												
Temp [°C]	microstructure					σ_{yi} [MPa]					σ_y [MPa]	a^{se} [-]
	PM	AY	FE ,PE	BA	MA	AY	FE, PE	BA	MA	PM		
950	63 %	37 %	0 %	0 %	0 %	53	38	38	38	38	43	18
580	63 %	37 %	0 %	0 %	0 %	96	147	301	417	106	102	18
560	63 %	33 %	4 %	0 %	0 %	96	157	322	439	150	133	17
410	63 %	0 %	4 %	33 %	0 %	98	197	435	560	244	304	14
20	63 %	0 %	4 %	33 %	0 %	200	258	490	620	350	411	15

CMM 6 (T_{max} = 1150 °C, t_{85} = 35 s)												
Temp [°C]	microstructure					σ_{yi} [MPa]					σ_y [MPa]	a^{se} [-]
	PM	AY	FE ,PE	BA	MA	AY	FE, PE	BA	MA	PM		
1150	31 %	69 %	0 %	0 %	0 %	29	14	14	14	15	25	21
580	31 %	69 %	0 %	0 %	0 %	96	147	301	417	106	99	21
560	31 %	61 %	8%	0 %	0 %	96	157	322	439	150	118	20
410	31 %	0 %	8%	61 %	0 %	98	197	435	560	244	356	14
20	31 %	0 %	8%	61 %	0 %	200	258	490	620	350	427	14

CMM 7 (T_{max} = 1350 °C, t_{85} = 35 s)												
Temp [°C]	microstructure					σ_{yi} [MPa]					σ_y [MPa]	a^{se} [-]
	PM	AY	FE ,PE	BA	MA	AY	FE, PE	BA	MA	PM		
1350	0	100 %	0 %	0 %	0 %	5	5	5	5	5	5	24
580	0	100 %	0 %	0 %	0 %	96	147	301	417	106	96	24
560	0	88%	12 %	0 %	0 %	96	157	321,6	439	150	103	23
410	0	0 %	12 %	88%	0 %	98	1967	435	561	244	406	14
20	0	0 %	12 %	88%	0 %	200	258	490	620	350	462	14

S355 of single-pass butt welds

CMM 2 (T_{max} = 826 oC, t_{85} = 32 s)												
Temp [°C]	microstructure					σ_{yi} [MPa]					σ_y [MPa]	a^{se} [-]
	PM	AY	FE ,PE	BA	MA	AY	FE, PE	BA	MA	PM		
826	69 %	31 %	0 %	0 %	0 %	59	67	71	123	53	55	13
660	69 %	31 %	0 %	0 %	0 %	91	102	196	278	116	108	13
620	69 %	25 %	6 %	0 %	0 %	98	125	252	356	150	136	13
480	69 %	2 %	6 %	23 %	0 %	144	190	395	517	280	299	12
20	69 %	0 %	6 %	23 %	2 %	245	258	490	620	355	386	12

CMM 3 (T_{max} = 931 °C, t_{85} = 32 s)												
Temp [°C]	microstructure					σ_{yi} [MPa]					σ_y [MPa]	a^{se} [-]
	PM	AY	FE ,PE	BA	MA	AY	FE, PE	BA	MA	PM		
931	37 %	63 %	0 %	0 %	0 %	40	40	40	55	30	36	15
660	37 %	63 %	0 %	0 %	0 %	91	102	196	278	116	100	15
620	37 %	51 %	12 %	0 %	0 %	98	125	252	356	150	121	14
480	37 %	4 %	12 %	47 %	0 %	144	190	395	517	280	318	12
20	37 %	0 %	12 %	47 %	4 %	245	258	490	620	355	417	12

CMM 4 (T_{max} = 1276 °C, t_{85} = 32 s)												
Temp [°C]	microstructure					σ_{yi} [MPa]					σ_y [MPa]	a^{se} [-]
	PM	AY	FE ,PE	BA	MA	AY	FE, PE	BA	MA	PM		
1276	0 %	100 %	0 %	0 %	0 %	5	5	5	5	5	0	16
660	0 %	100 %	0 %	0 %	0 %	91	102	196	278	116	91	16
620	0 %	81 %	19 %	0 %	0 %	98	125	252	356	150	103	15
480	0 %	6 %	19 %	75 %	0 %	144	190	395	517	280	341	12
20	0 %	0 %	19 %	75 %	6 %	245	258	490	620	355	454	12

CMM 5 (T_{max} = 826 °C, t_{85} = 85 s)												
Temp [°C]	microstructure					σ_{yi} [MPa]					σ_y [MPa]	a^{se} [-]
	PM	AY	FE ,PE	BA	MA	AY	FE, PE	BA	MA	PM		
826	67 %	33 %	0 %	0 %	0 %	59	67	71	123	64	62	13
690	67 %	33 %	0 %	0 %	0 %	85	84	154	220	91	89	13
660	67 %	25 %	8%	0 %	0 %	91	102	196	278	116	108	13
620	67 %	23 %	10 %	0 %	0 %	98	125	252	356	150	136	13
500	67 %	2 %	10 %	21 %	0 %	121	187	140	505	261	226	12

CMM 6 (T_{max} = 931 °C, t_{85} = 85 s)												
Temp [°C]	microstructure					σ_{yi} [MPa]					σ_y [MPa]	a^{se} [-]
	PM	AY	FE ,PE	BA	MA	AY	FE, PE	BA	MA	PM		
931	33 %	67 %	0 %	0 %	0 %	40	40	40	55	51	44	15
690	33 %	67 %	0 %	0 %	0 %	85	84	154	220	91	87	15
660	33 %	51 %	16 %	0 %	0 %	91	102	196	278	116	101	14
620	33 %	47 %	20 %	0 %	0 %	98	125	252	356	150	121	14
500	33 %	4 %	20 %	43 %	1 %	121	187	140	505	261	191	12

CMM 7 (T_{max} = 1276 °C, t_{85} = 85 s)												
Temp [°C]	microstructure					σ_{yi} [MPa]					σ_y [MPa]	a^{se} [-]
	PM	AY	FE ,PE	BA	MA	AY	FE, PE	BA	MA	PM		
1276	0 %	100 %	0 %	0 %	0 %	5	5	5	5	171	5	16
690	0 %	100 %	0 %	0 %	0 %	85	84	154	220	91	85	16
660	0 %	75 %	25 %	0 %	0 %	91	102	196	278	116	93	15
620	0 %	69 %	31 %	0 %	0 %	98	125	252	356	150	107	15
500	0 %	0 %	31 %	68%	1 %	121	187	140	505	261	158	12

S355 of project *HFH-Simulation*

CMM 2 (T_{max} = 812 °C, t_{85} = 4 s)												
Temp [°C]	microstructure					σ_{yi} [MPa]					σ_y [MPa]	a^{se} [-]
	PM	FE, PE	BA	MA	AY	PM	FE, PE	BA	MA	AY		
812	66 %	0 %	0 %	0 %	34 %	100	79	42	120	62	87	13
713	66 %	0 %	0 %	0 %	34 %	153	86	73	264	81	128	13
420	66 %	0 %	0 %	0 %	34 %	292	218	237	783	136	239	13
200	66 %	0 %	0 %	31 %	3 %	353	285	271	880	200	514	12
20	66 %	0 %	0 %	31 %	3 %	400	431	277	1008	245	586	12

CMM 3 (T_{max} = 877 °C, t_{85} = 4 s)												
Temp [°C]	microstructure					σ_{yi} [MPa]					σ_y [MPa]	a^{se} [-]
	PM	FE, PE	BA	MA	AY	PM	FE, PE	BA	MA	AY		
877	33 %	0 %	0 %	0 %	67 %	65	57	29	78	49	55	15
713	33 %	0 %	0 %	0 %	67 %	153	86	73	264	81	104	15
420	33 %	0 %	0 %	0 %	67 %	292	218	237	783	136	188	15
200	33 %	0 %	0 %	62 %	5 %	353	285	271	880	200	670	12
20	33 %	0 %	0 %	62 %	5 %	400	431	277	1008	245	766	12

CMM 4 (T_{max} = 1205 °C, t_{85} = 4 s)												
Temp [°C]	microstructure					σ_{yi} [MPa]					σ_y [MPa]	a^{se} [-]
	PM	FE, PE	BA	MA	AY	PM	FE, PE	BA	MA	AY		
1205	0 %	0 %	0 %	0 %	100 %	5	5	5	5	5	5	16
713	0 %	0 %	0 %	0 %	100 %	153	86	73	264	81	81	16
420	0 %	0 %	0 %	0 %	100 %	292	218	237	783	136	136	16
200	0 %	0 %	0 %	92 %	8%	353	285	271	880	200	826	12
20	0 %	0 %	0 %	92 %	8%	400	431	277	1008	245	947	12

S690 of project *HFH-Simulation*

CMM 2 (T_{max} = 817 °C, t_{85} = 4 s)												
Temp	microstructure					σ_{yi} [MPa]					σ_y	a^{se}
[°C]	PM	FE, PE	BA	MA	AY	PM	FE, PE	BA	MA	AY	[MPa]	[-]
817	63 %	0 %	0 %	0 %	37 %	97	85	58	113	61	84	13
719	63 %	0 %	0 %	0 %	37 %	150	94	101	246	79	124	13
431	63 %	0 %	0 %	0 %	37 %	594	237	332	750	134	424	13
200	63 %	0 %	0 %	34 %	3 %	676	312	386	853	200	722	12
20	63 %	0 %	0 %	34 %	3 %	734	472	394	977	245	802	12

CMM 3 (T_{max} = 882 °C, t_{85} = 4 s)												
Temp	microstructure					σ_{yi} [MPa]					σ_y	a^{se}
[°C]	PM	FE, PE	BA	MA	AY	PM	FE, PE	BA	MA	AY	[MPa]	[-]
850	31 %	0 %	0 %	0 %	69 %	62	61	40	73	48	53	15
719	31 %	0 %	0 %	0 %	69 %	150	94	101	246	79	101	15
431	31 %	0 %	0 %	0 %	69 %	594	237	332	750	134	277	15
200	31 %	0 %	0 %	63 %	6 %	676	312	386	853	200	762	12
20	31 %	0 %	0 %	63 %	6 %	734	472	394	977	245	861	12

CMM 4 (T_{max} = 1208 °C, t_{85} = 4 s)												
Temp	microstructure					σ_{yi} [MPa]					σ_y	a^{se}
[°C]	PM	FE, PE	BA	MA	AY	PM	FE, PE	BA	MA	AY	[MPa]	[-]
1208	0 %	0 %	0 %	0 %	100 %	26	12	8	13	11	11	16
719	0 %	0 %	0 %	0 %	100 %	150	94	101	246	79	79	16
431	0 %	0 %	0 %	0 %	100 %	594	237	332	750	134	134	16
200	0 %	0 %	0 %	92 %	8%	676	312	386	853	200	801	12
20	0 %	0 %	0 %	92 %	8%	734	472	394	977	245	918	12

S960 of project *HFH-Simulation*

CMM 2 (T_{max} = 866 °C, t_{85} = 4 s)												
Temp [°C]	microstructure					σ_{yi} [MPa]					σ_y [MPa]	a^{se} [-]
	PM	FE, PE	BA	MA	AY	PM	FE, PE	BA	MA	AY		
866	63 %	0 %	0 %	0 %	37 %	86	71	62	87	51	73	13
719	63 %	0 %	0 %	0 %	37 %	208	99	139	259	79	161	13
431	63 %	0 %	0 %	0 %	37 %	826	251	457	787	134	570	13
200	63 %	0 %	0 %	34 %	3 %	940	331	531	895	200	903	12
20	63 %	0 %	0 %	34 %	3 %	1021	500	542	1025	245	999	12

CMM 3 (T_{max} = 964 °C, t_{85} = 4 s)												
Temp [°C]	microstructure					σ_{yi} [MPa]					σ_y [MPa]	a^{se} [-]
	PM	FE, PE	BA	MA	AY	PM	FE, PE	BA	MA	AY		
915	31 %	0 %	0 %	0 %	69 %	42	33	24	23	33	36	15
719	31 %	0 %	0 %	0 %	69 %	208	99	139	259	79	119	15
431	31 %	0 %	0 %	0 %	69 %	826	251	457	787	134	349	15
200	31 %	0 %	0 %	63 %	6 %	940	331	531	895	200	871	12
20	31 %	0 %	0 %	63 %	6 %	1021	500	542	1025	245	981	12

CMM 4 (T_{max} = 1257 °C, t_{85} = 4 s)												
Temp [°C]	microstructure					σ_{yi} [MPa]					σ_y [MPa]	a^{se} [-]
	PM	FE, PE	BA	MA	AY	PM	FE, PE	BA	MA	AY		
1257	0 %	0 %	0 %	0 %	100 %	5	5	5	5	5	5	16
719	0 %	0 %	0 %	0 %	100 %	208	99	139	259	79	79	16
431	0 %	0 %	0 %	0 %	100 %	826	251	457	787	134	134	16
200	0 %	0 %	0 %	92 %	8%	940	331	531	895	200	839	12
20	0 %	0 %	0 %	92 %	8%	1021	500	542	1025	245	963	12

Appendix B – FE Background

Shape Functions of Applied Finite Elements

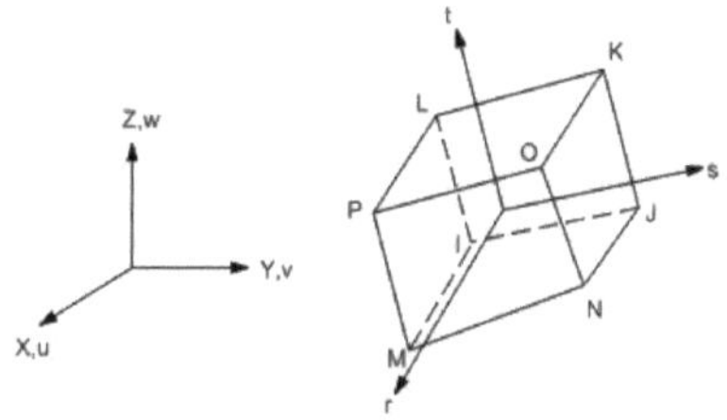

shape functions of *ANSYS Solid70*

$$T = \frac{1}{8} \cdot \{T_I \cdot (1-s) \cdot (1-t) \cdot (1-r) + T_J \cdot (1-s) \cdot (1-t) \cdot (1-r)$$
$$+ T_K \cdot (1-s) \cdot (1-t) \cdot (1-r) + T_L \cdot (1-s) \cdot (1-t) \cdot (1-r)$$
$$+ T_M \cdot (1-s) \cdot (1-t) \cdot (1-r) + T_N \cdot (1-s) \cdot (1-t) \cdot (1-r)$$
$$+ T_O \cdot (1-s) \cdot (1-t) \cdot (1-r) + T_P \cdot (1-s) \cdot (1-t) \cdot (1-r)\}$$

shape functions of *ANSYS Solid185*

(analogous for *LS-DYNA* 8-node hexahedron solid elements*)*

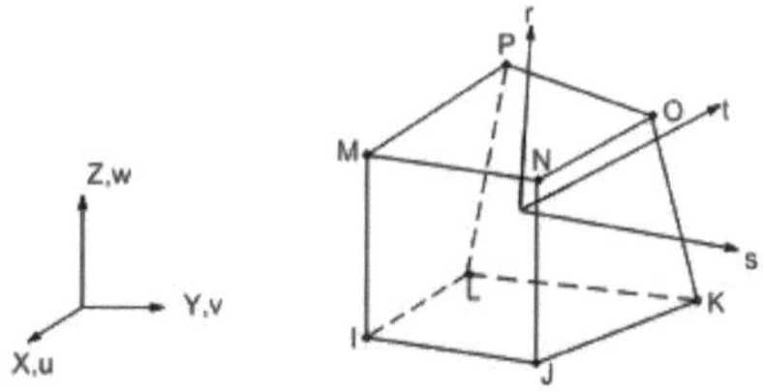

$$u = \frac{1}{8} \cdot \{u_I \cdot (1-s) \cdot (1-t) \cdot (1-r) + u_J \cdot (1-s) \cdot (1-t) \cdot (1-r)$$
$$+ u_K \cdot (1-s) \cdot (1-t) \cdot (1-r) + u_L \cdot (1-s) \cdot (1-t) \cdot (1-r)$$
$$+ u_M \cdot (1-s) \cdot (1-t) \cdot (1-r) + u_N \cdot (1-s) \cdot (1-t) \cdot (1-r)$$
$$+ u_O \cdot (1-s) \cdot (1-t) \cdot (1-r) + u_P \cdot (1-s) \cdot (1-t) \cdot (1-r)\}$$

$v = \frac{1}{8} \cdot \{v_I \cdot (1-s)$ ... (analogous to u)

$w = \frac{1}{8} \cdot \{w_I \cdot (1-s)$ ... (analogous to u)

Appendix C – Statistics

Cowper-Symonds - S355 parent material

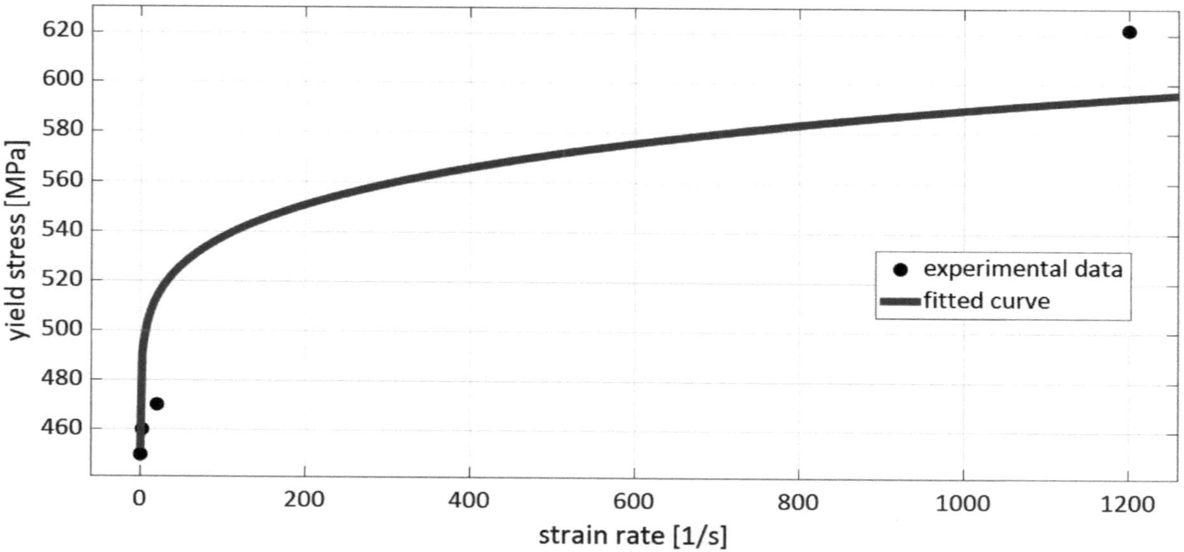

General model: f(x) = (x/D)^(1/q)*450+450

Coefficients (with 95 % confidence bounds): D = 3.535e+05 (-8.409e+05, 1.548e+06), q = 5 (fixed at bound)

Goodness of fit: SSE = 3584, R-square: 0.8197, Adjusted R-square: 0.8197, RMSE: 34.56

Cowper-Symonds – S690 parent material

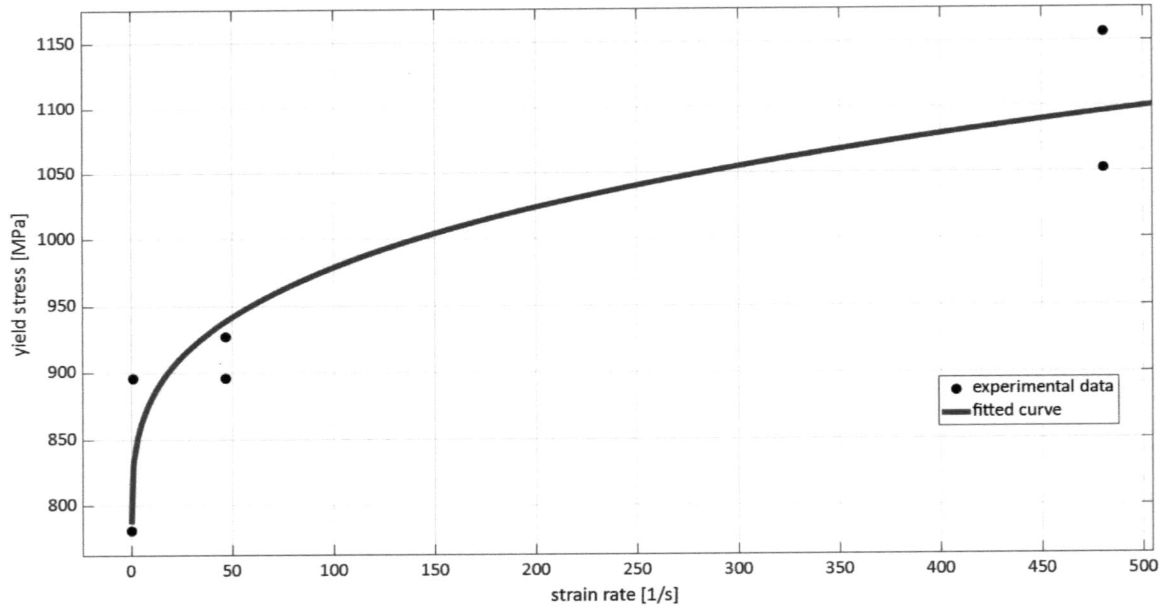

General model: f(x) = (x/D)^(1/q)*780+780

Coefficients (with 95 % confidence bounds): D = 1.036e+04 (-2.605e+04, 4.677e+04), q = 3.394 (0.2268, 6.561)

Goodness of fit: SSE = 1.204e+04, R-square = 0.8629, Adjusted, R-square = 0.8286, RMSE = 54.86

Cowper-Symonds – S960 parent material

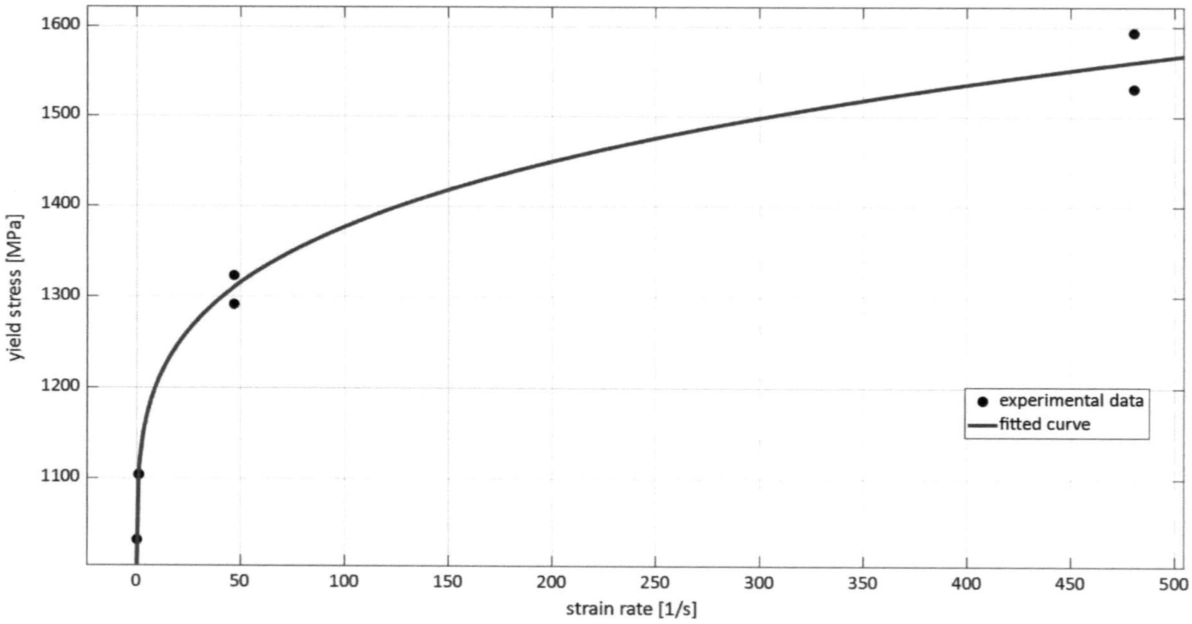

General model: f(x) = (x/D)^(1/q)*952+952

Coefficients (with 95 % confidence bounds): D = 4156 (1556, 6756), q = 4.125 (3.338, 4.912)

Goodness of fit: SSE = 3226, R-square = 0.9888, Adjusted R-square: 0.9865, RMSE: 25.4

Cowper-Symonds – S355 Heat Affected Zone

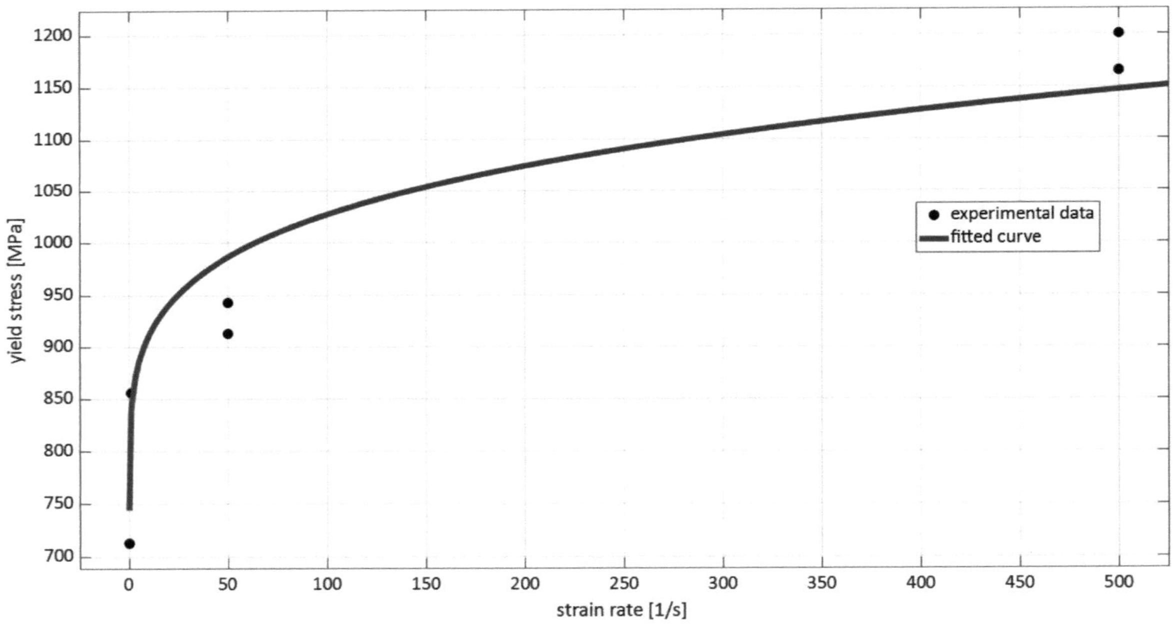

General model: f(x) = (x/D)^(1/q)*713+713

Coefficients (with 95 % confidence bounds): D = 5993 (912.5, 1.107e+04), q = 5 (fixed at bound)

Goodness of fit: SSE = 1.181e+04, R-square = 0.932, Adjusted R-square: 0.932, RMSE: 48.6

Cowper-Symonds – S690 Heat Affected Zone

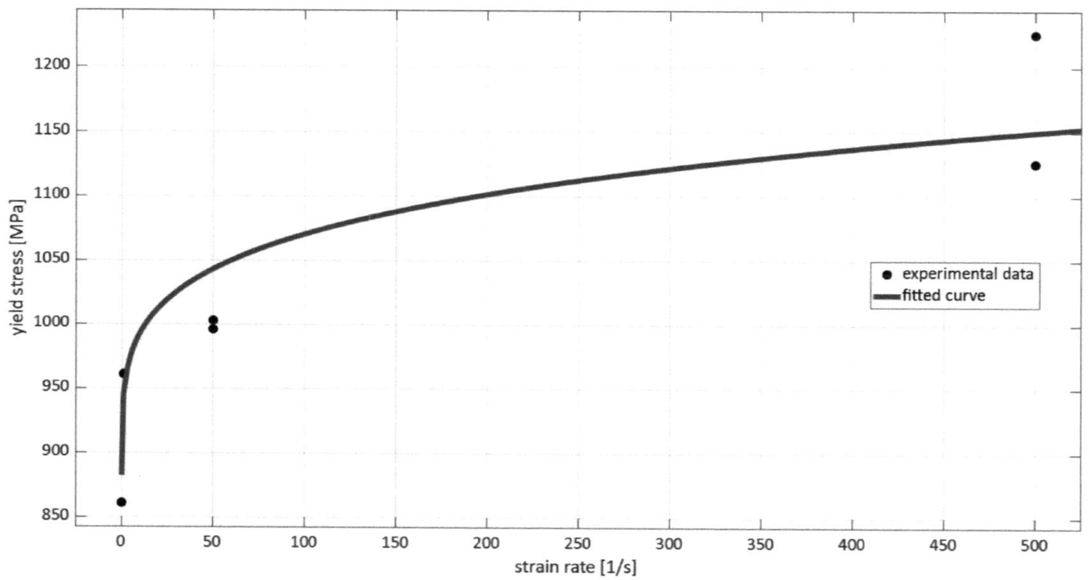

General model: f(x) = (x/D)^(1/q)*861+861

Coefficients (with 95 % confidence bounds): D = 1.174e+05 (-2.638e+04, 2.612e+05), q = 5 (fixed at bound)

Goodness of fit: SSE = 1.094e+04, R-square = 0.868, Adjusted R-square: 0.868, RMSE: 46.77

Cowper-Symonds – S960 Heat Affected Zone

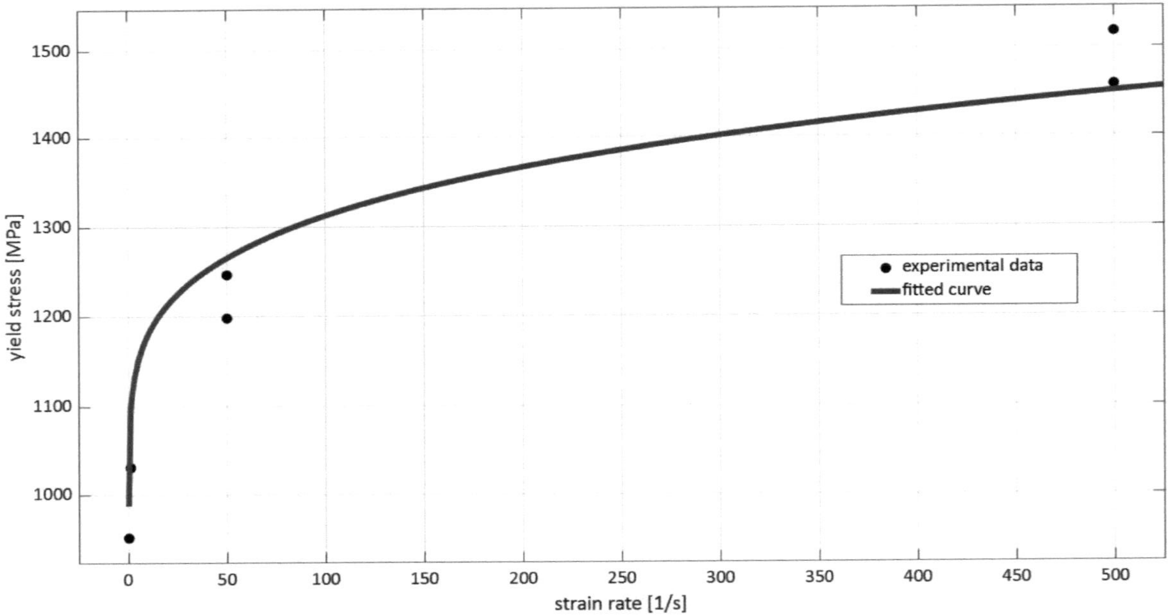

General model: f(x) = (x/D)^(1/q)*952+952

Coefficients (with 95 % confidence bounds): D = 1.283e+04 (2076, 2.358e+04), q = 5 (fixed at bound)

Goodness of fit: SSE = 1.518e+04, R-square = 0.9399, Adjusted R-square: 0.9399, RMSE: 55.11

BERICHTE ZUM STAHL- UND LEICHTBAU (ISSN 2198-7912)

Versuchsanstalt für Stahl, Holz und Steine, Stahl- und Leichtbau
Karlsruher Institut für Technologie (KIT)

Eine Übersicht der Berichte der Versuchsanstalt für Stahl, Holz und Steine ab dem Jahr 1963 finden Sie unter folgender URL: http://stahl.vaka.kit.edu/berichte.php

Band 1 **OLIVER FLEISCHER**
Axial beanspruchte K-Knoten aus dünnwandigen Rechteckhohlprofilen. 2014
ISBN 978-3-7315-0190-9

Band 2 **THOMAS REINKE**
Tragverhalten von biegebeanspruchten Stahlmasten mit polygonalen Querschnitten. 2015
ISBN 978-3-7315-0398-9

Band 3 **ROBIN MARC PLUM**
Fatigue crack detection on structural steel members by using ultrasound excited thermography. Erkennung von Ermüdungsrissen in Stahlbauteilen durch ultraschallangeregte Thermografie. 2015
ISBN 978-3-7315-0417-7

Band 4 **TIM ZINKE**
Nachhaltigkeit von Infrastrukturbauwerken – Ganzheitliche Bewertung von Autobahnbrücken unter besonderer Berücksichtigung externer Effekte. 2016
ISBN 978-3-7315-0509-9

Band 5 **MAX JONAS SPANNAUS**
Bemessung von Erzeugnissen aus Stahlguss unter vorwiegend ruhender Beanspruchung. 2016
ISBN 978-3-7315-0560-0

Band 6 **MATTHIAS FRIEDRICH ALBIEZ**
Zur statischen Tragfähigkeit geklebter Kreishohlprofilverbindungen im Stahlbau. 2016
ISBN 978-3-7315-0561-7

Band 7 **ANDREAS LIPP**
Kreishohlprofil-X-Knoten aus nichtrostenden Stählen unter Axialbeanspruchung. 2016
ISBN 978-3-7315-0569-3

Band 8 **PAUL DARIO TOASA CAIZA**
Consideration of runouts by the evaluation of fatigue experiments. 2019
ISBN 978-3-7315-0900-4

Band 9 **JENNIFER C. HRABOWSKI**
Ermüdungsverhalten von Schweißverbindungen aus höchstfestem Stahl im Kurzzeitfestigkeitsbereich. 2019
ISBN 978-3-7315-0931-8

Die Bände sind unter www.ksp.kit.edu als PDF frei verfügbar oder als Druckausgabe bestellbar.

Band 10 **STEFAN RACK**
Formadaptive Tensegrity-Strukturen – Ein Beitrag zur numerischen Simulation. 2019
ISBN 978-3-7315-0574-7

Band 11 **ACHIM STRUVE**
Analysis of a Rotatable Wind Turbine Tower by means of Aero-Servo-Elastic Load Simulations. 2021
ISBN 978-3-7315-1045-1

Band 12 **PHILIPP WEIDNER**
Zum Ermüdungsverhalten einseitig geschweißter Stumpfstöße von Kreishohlprofilen. 2021
ISBN 978-3-7315-1017-8

Band 13 **STEFANOS GKATZOGIANNIS**
Finite Element Simulation of Residual Stresses from Welding and High Frequency Hammer Peening. 2021
ISBN 978-3-7315-1066-6